Science, Technology and Innovation Studies

Series editors
Leonid Gokhberg
Moscow, Russia

Dirk Meissner
Moscow, Russia

Science, technology and innovation (STI) studies are interrelated, as are STI policies and policy studies. This series of books aims to contribute to improved understanding of these interrelations. Their importance has become more widely recognized, as the role of innovation in driving economic development and fostering societal welfare has become almost conventional wisdom. Interdisciplinary in coverage, the series focuses on the links between STI, business, and the broader economy and society. The series includes conceptual and empirical contributions, which aim to extend our theoretical grasp while offering practical relevance. Relevant topics include the economic and social impacts of STI, STI policy design and implementation, technology and innovation management, entrepreneurship (and related policies), foresight studies, and analysis of emerging technologies. The series is addressed to professionals in research and teaching, consultancies and industry, government and international organizations.

More information about this series at http://www.springer.com/series/13398

Pierre-Bruno Ruffini

Science and Diplomacy

A New Dimension
of International Relations

 Springer

Pierre-Bruno Ruffini
Faculty of International Affairs
University of Le Havre
Le Havre, France

Translation from the French language edition: Science et diplomatie. Une nouvelle dimension des relations internationales, 978-2-84924-391-6 © Editions du Cygne, Paris, 2015. All
Rights Reserved.

Science, Technology and Innovation Studies
ISBN 978-3-319-55103-6   ISBN 978-3-319-55104-3   (eBook)
DOI 10.1007/978-3-319-55104-3

Library of Congress Control Number: 2017939354

Printed on acid-free paper

This Springer imprint is published by Springer Nature
The registered company is Springer International Publishing AG
The registered company address is: Gewerbestrasse 11, 6330 Cham, Switzerland

Acknowledgement

The author thanks the science counselors and attachés of diplomatic networks for their assistance in collecting information and for shared reflections on science diplomacy, especially Véronique Briquet-Laugier (SST, French Embassy in India), Kirill Bykov (Ministry of Foreign Affairs of the Russian Federation), Philippe Carlevan (SST, French Embassy in Canada), Alain Derevier (Permanent Representation of France to the FAO, Rome), Jean-Marie Freyssinet (SST, French Embassy in Russia), Candy Green (Economic-Environment, Science, Technology & Health Section, State Department, USA), Erik Hall (Economic-Environment, Science, Technology & Health Section, State Department, USA), Matthew Houlihan (SIN, UK), Sébastien Hug (Federal Department of Economic Affairs, Education and Research, Switzerland), Ichiro Ikeda (Ministry of Foreign Affairs of Japan), Emmanuel Kamarianakis (Canadian Trade Commissioner Service), Alison Mac Ewen (SIN, UK), Philippe Martineau (SST, French Embassy in China), Vladimir Majer (Section of the Deputy Prime Minister and Minister for Science, Research and Innovation of the Czech Republic), Klaus Matthes (Ministry of Foreign Affairs of Germany), Laura Nuccilli (SIN, UK), Oleg Ossipov (Ministry of Foreign Affairs of the Russian Federation), Min-Hà Pham (SST, French Embassy in the USA), John Picard (Canadian Trade Commissioner Service), Serge Plattard (SST, French Embassy in the UK), Petr Reimer (Ministry of Foreign Affairs of the Czech Republic), Florence Rivière-Bourhis (SST, French Embassy in Japan), Stéphane Roy (SST, French Embassy in Germany), Kirsten Schultz (Economic-Environment, Science, Technology & Health Section, State Department, USA), Szilvia Szántó (Ministry of Foreign Affairs of Hungary), Annick Suzor-Weiner (SST, French Embassy in the USA), Norimasa Takeda (Ministry of Foreign Affairs of Japan), Ruth Theus Baldassarre (Federal Department of Foreign Affairs, Switzerland), Matthieu Weiss (SST, French Embassy in Germany) and Jun Yin (MOST, China).

The author also thanks for their time and advice during the preparation of this book: Stefano Ajola (Economic Service, German Embassy in Italy), Jean-Claude Arditti (AVRIST), Stefano Baldi (Ministry of Foreign Affairs of Italy), Nicolas Bériot (ONERC), Catherine Bréchignac (Academy of Sciences), Roberto Cantone (Ministry of Foreign Affairs of Italy), Lucien Chabason (IDDRI), Jean-Luc Clément (MESR), Michel Colombier (IDDRI), Yves Frénot (French Polar Institute),

François Gave (MAE), Mireille Guigaz (MAE), Jennifer Heurley (MAE), Didier Hoffschir (MESR), Jean-Charles Hourcade (CIRED), Marcel Jouve (MAE), Jean Jouzel (IPCC), Anne Larigauderie (National Museum of Natural History, Paris), Hervé Le Treut (IPSL), Philippe Martinet (MAE), David Musial (Ministry of Foreign Affairs of Germany), Édith Ravaux (MAE), Scott Stone (Cabinet Hunton& Williams), Doug Trappett (Ministry of Foreign Affairs of Australia) and Paul Watkinson (Ministry of Ecology, Sustainable Development and Energy).

Contents

List of Abbreviations

AVRIST	Association pour la Valorisation des Relations Internationales Scientifiques et Techniques (Paris)
CIRED	International Research Center on Environment and Development (Paris)
CNRS	National Center for Scientific Research (France)
IDDRI	Institute for Sustainable Development and International Relations (Paris)
IPCC	Intergovernmental Panel on Climate Change
IPSL	Institute Pierre et Simon Laplace (Paris)
MAE	Ministry of Foreign Affairs (France)[1]
MESR	Ministry of Higher Education and Research (France)[2]
MOST	Ministry of Science and Technology (China)
ONERC	Observatoire National sur les Effets du Réchauffement Climatique (France)
SIN	Science and Innovation Network (UK)
SST	Service for Science and Technology

---

[1]Ministry of Foreign Affairs and International Development since 2014.

[2]Secretariat of State for Higher Education and Research since 2013.

Introduction

Geneva, 19 November 1985. Mikhail Gorbachev and Ronald Reagan met for the first time. During this summit mainly devoted to disarmament, the Soviet leader proposed to his American counterpart an ambitious program of research and experimentation on a subject on which scientists of his country had been devoting much attention for years: verifying the scientific and technological feasibility of nuclear fusion as a new source of energy. Very consistent with the detente and cooling of tensions in the post-Cold War world, this proposal for a peaceful use of nuclear energy had considerable symbolic meaning. More importantly, it had huge prospects for controlling the process of nuclear fusion in order to produce clean and virtually limitless energy from abundant elements in nature, provided a long-term scientific investment was made. The following year, the United States, the European Union and Japan agreed to join the Soviet Union and conduct this program, whose economic reach, if successful, would be considerable: the ITER (International Thermonuclear Experimental Reactor) was born. Later, China, India and South Korea joined the adventure. On 17 November 2010, the foundation stone of the experimental reactor was laid in France on the Cadarache site.

Cairo, 4 June 2009. Eight years after the September 11 attacks, President Obama delivered a speech at Cairo University that marked a strong shift in American policy toward the Arab-Muslim world. This inspired, peace seeking and constructive speech was a way of reaching out to countries in order to show them that America could speak in a language other than that of force. The American President announced the creation of a new fund to support technological development. He promised the opening of centers of excellence in Africa, the Middle East and Southeast Asia. He announced the dispatch of science envoys to Muslim countries. Just a few weeks later, Bruce Albert, twice president of the National Academy of Sciences, Elias Zerhouni, who directed the National Institutes of Health, and Ahmed Zouheil, Nobel laureate in chemistry, were in the field. These renowned American researchers examined the possibilities of cooperation in all scientific and technical fields. They recommended the establishment of three centers of scientific excellence on the themes of water, climate and political science.

© Springer International Publishing AG 2017
P.-B. Ruffini, *Science and Diplomacy*, Science, Technology and Innovation Studies,
DOI 10.1007/978-3-319-55104-3_1

Stockholm, 27 September 2013. After 4 days of talks and a final night of intense discussion in the large conference room of Münchenbryggeriet, the 12th session of the Scientific Working Group of the Intergovernmental Panel on Climate Change (IPCC) came to its end. More than 400 people attended the meeting, including representatives of 116 governments and UN (IPCC) climate experts and many observers from various organizations. Official delegates and scientific authors of the "report for policy makers" were to approve the wording at the end of a long line-by-line discussion. This text synthesized in 30 pages the present state of knowledge on climate, the origin of its irregularities and its future prospects. Entitled *Climate Change 2013—The Physical Science Basis*, this contribution was one of the pieces of the puzzle to be completed a few months later by the other two working groups of the IPCC which were dedicated to the impact of climate change and to its mitigation, and culminating in 2014 with the publication of the Fifth Assessment Report.

Addressing the relationship between science and diplomacy is addressing *science diplomacy*, which is situated, to start with, in the particular field of international relations where the interests of science and those of foreign policy intersect. We have given three examples that illustrate its different aspects. ITER is a global project that transcends the specific interests of each country involved in it. It illustrates how diplomacy works in support of science, which a pioneering study in 2010 by the Royal Society and the American Association for the Advancement of Science (AAAS) called "diplomacy for science".[1] At the origin of the ITER project, there was an intuition emanating from the world of science: harnessing the considerable energy from atoms coming close enough to each other to join in the process of fusion, such an idea could only sprout in the minds of specialists. But if physicists have dreamed about it, it is the politicians who made the project happen. The Russian scientist Evegeniy Velikhov, president of the scientific council of ITER, reflected in these terms: "We knew that only a broad international program could allow us to demonstrate the scientific and technological feasibility of fusion energy. In 1985, Gorbachev, who had extensively discussed the project with President Mitterrand during his first visit to France, a month before the summit in Geneva, proposed it to Reagan. And that's how it all really started".[2] Initial thoughts on the ITER project in the 1970s would have come to nothing without the strong and sustained commitment of the leaders of the most powerful countries and the persistent efforts of diplomats, who had to solve thorny questions regarding the choice of the host country and the funding of the experimental reactor, until the final signing of the agreement at the Elysée Palace on November 21, 2006.[3]

The Cairo speech illustrates another dimension of science diplomacy: that of science as a facilitator or vanguard of diplomacy, which the Royal Society-AAAS study cited above named "science for diplomacy". When political tensions between

---

[1] We will refer to this text as the "Royal Society-AAAS report".

[2] http://www.iter.org/actualites/501.

[3] The cost of construction of the reactor is estimated at around 19 billion euros. The operational phase is expected to start in the late 2020s.

countries do not allow traditional diplomacy to take place, scientific relations can be used to maintain or restore the links. In the Cairo speech, President Obama made announcements on scholarships to be granted to students from the Arab-Muslim world and on the launch of a new fund to support technological development in these countries. Pursuing an objective of national security in the long term, the American President emphasized the role of science to correct the negative image left by America in this part of the world as a result of the war against Iraq.[4]

The third example illustrates the way science can inform diplomatic choices. The climate issue is a science-intensive one. By performing a thorough survey of knowledge on climate and refining the diagnosis over its successive reports, the collective expertise conducted by the IPCC provides the elements which are needed by diplomats involved in negotiations that take place regularly in the framework of the "Climate Convention". This input of science in diplomacy manifests itself most strongly in the method at work in the IPCC: in Stockholm, government representatives have weighed every word of the conclusions put on the table by scientists, not to challenge the content but to adapt formulation to the requirements of diplomatic exercises that would come later. After their last night of effort, they agreed with the group of experts on the final summary for policy makers. Official delegates, many of them future negotiators in international forums on behalf of their country, left Stockholm after having made their own scientific statement based on the reports they had approved. This approach forcefully illustrates the role of science to feed and legitimize diplomatic decisions. It falls under the third area of the three main science-diplomacy relations identified by the Royal Society and AAAS, that of "science in diplomacy".

Science Diplomacy in a Rapidly Changing World

At the crossroads of the two fields, that of knowledge production and that of foreign affairs, science diplomacy is in the air. But this trend owes nothing to chance. The world of science and that of diplomacy have for decades been subject to their own developments which have encouraged closer relations between them. It is best to start by pointing out this context.

The Changing Global Landscape of Science

What is science? The sociologist Robert Merton clarified the complexity of what the term covers: science designates at the same time methods to produce and certify knowledge, the stock of the resulting accumulated knowledge, and a set of cultural values and conducts that govern scientific activity (1942, p. 268). Scientific research designates more specifically the production process of new knowledge. But it is important to agree on fields of knowledge that are within science. A

---

[4]This diplomatic shift was also pursuing another goal: counteract the depletion of intake of students and young researchers from developing countries which followed the tightening of immigration policy under Patriot Act of 26 October 2001, and which was detrimental to research activity in the United States.

common practice in France or in Anglo-Saxon countries is that science is restricted to natural sciences only.[5] We retain here the broad sense of the word "science", which covers the entire spectrum embracing exact and natural sciences as well as social sciences and humanities. There are at least two reasons for this. A matter of principle, first of all, is that the approaches of researchers in almost all disciplines share the common feature of moving back and forth between the observation of facts and the construction of theories and models to interpret them. In this sense, the sociologist or the economist is no less "scientific" than the physicist or the biologist. Another reason is institutional: in public policy, funding and other support for research, project selection procedures or methods of evaluation of results are part of a comprehensive approach in which the specifics of each discipline does not occupy the foreground.

The global landscape of science is changing. We retain five trends:

- **The weight of research and development (R&D) in the global economy is rising**. In 2013, total expenditures on R&D in the world stood at about 1478 billion US dollars. These expenses totaled just over 500 billion in 1996 (UNESCO 2015, p. 24). This dynamism of investment in R&D becomes fully apparent when compared to that of wealth creation as a whole: in the period, global spending on R&D grew faster than world GDP, and is about 1.7% of that total today. The robust growth in production of new knowledge is also evident in the changing of the global research workforce, which increased from about four million in 1995 to nearly 7.8 million in 2013 (*ibid.*, p. 33). The same steady growth occurs when considering the number of scientific publications: the number of articles published in internationally recognized academic journals has grown from about 460,200 in 1988 to about 828,000 in 2011(National Science Board 2012).
- **Scientific research is increasingly international**. Collaborations between researchers from different countries lead to co-publications, whose share in all international publications has risen from 13.2% in 2000 to 19.2% in 2013 (National Science Board 2016).[6] This increasing internationalization is also measured by the number of citations which indicates the usefulness of publications to other researchers: in most countries, citations of articles in the international scientific literature have increased at the expense of citations of purely domestic articles.
- One aspect of the internationalization of research is due to the development of *Big Science*, which is the concentration of budgets and staff researchers on very large scale experimental equipments. *Big Science* emerged during World War II

---

[5]J.-J. Salomon (2006, p. 237) notes that "on the European continent, the very concept of science (*Wissenschaft* in Germany or *nauka* in Russia) has always included the social sciences and humanities", while "in the Anglo-Saxon world, the policy of natural sciences is separate from that which affects the social sciences...".

[6]These percentages relate to publications in science and engineering.

in close relation with the design of nuclear weapons. Today it requires considerable funding through international organizations or consortia in areas such as fundamental physics (Large Hadron Collider at the European Organization for Nuclear Research), astronomy and space (Hubble Space Telescope, International Space Station, Cassini mission to Saturn), or life sciences (Human Genome complete DNA sequencing Project—HGP).

– **The center of gravity of global production of knowledge is shifting**. The entry of the Asian continent onto the global research stage is the most spectacular event of the beginning of the third millennium. First of all, this shift is visible through the global distribution of R&D expenditures. In 2009, for the first time, total R&D in the region has exceeded 400 billion US dollars and is on a par with that of the United States. With R&D spending increasing at an annual rate of 20% since 1997, China has been the main architect of the overall vitality of the region, whose share of world expenditures on R&D increased from 24 to 42% between 1996 and 2013. Over the period, the share of North America (United States, Canada and Mexico) decreased from 40 to 30% and that of the European Union from 31 to 23%. The erosion of the relative contribution of the United States and of Europe to global research is also evidenced by the indicator of the number of publications. In 1995, the United States and Europe together accounted for 69% of world scientific production. In 2014, their share had fallen to 65% while that of Asia had risen from 14 to 40% (of which 20 percentage points for China).[7] Finally, China's share in the total research workforce amounted to 19% in 2013, whereas it was less than 14% in 2002. This is almost as much as the share of the European Union (22%) but higher than that of the United States (17%). This proliferation of hubs of expertise changes the mapping of international migrations of researchers: a rebalancing is under way, and there is today less talk about "brain drain" and more about "brain circulation".

– Lastly, many challenges the international community has to face are science-related and technology-driven. These **global issues** relate to human health (access to water and food resources, protection against diseases, especially infectious ones...), security concerns (fight against the proliferation of weapons of mass destruction, security of energy supply, security of digital information networks...) or the quality of the environment (preserving biodiversity, combating the negative effects of climate change or ocean pollution...). These challenges affect the future of mankind and no country can hope to tackle them alone. They contribute to an increased globalization of science and a renewal of diplomatic activity.

---

[7]Articles co written by authors from more than one region are accounted for in each of them, therefore the total of regional shares exceeds 100%. All figures in this paragraph are from UNESCO (2015), *op.cit.*

New Ways of Diplomacy

Diplomacy is the use of dialogue, negotiation and representation in international relations. It refers to actions and means other than the use of force or coercion, by which a country seeks to defend and promote its interests and values in its relations with other countries. Diplomacy is a profession, a career and an art all at the same time, enabling it to find arrangements between competing interests and avoiding confrontations (Delcorde 2009). It contributes to the implementation of foreign policy, with which it should not be confused.[8]

Embassies and consulates that a country deploys abroad are key components of its diplomatic infrastructure. The Vienna Convention defines the main functions of a diplomatic mission of a sending state in the receiving state: representing the sending state to protect its interests and those of its nationals; ascertaining conditions and developments in the receiving state, and negotiating with the government of the receiving state; and finally, promoting friendly relations and developing economic, cultural and scientific relations with the receiving state.[9] Note here that the development of bilateral scientific relations is explicitly mentioned among the duties of a diplomatic mission.

Diplomacy stems from and expresses national sovereignty on the international stage. It is traditionally the exclusive privilege of sovereign states.[10] In this sense, diplomacy was born with the modern systems of states that emerged from the Treaty of Westphalia (1648) and the organization of international relations that has resulted. This model has dominated international relations for nearly five centuries. Today's diplomacy stands out in many ways because of the major changes it has undergone. We identify four:

- **Multilateralization**. Adding to the usual care of state-to-state relations (bilateral diplomacy) is the involvement of national diplomacy in issues that affect several countries at once, or the community of nations as a whole. This multilateral diplomacy took off after World War II with the rise of the United Nations system: multilateral diplomacy is typically the "diplomacy of international organizations and international conferences" (Delcorde 2005, pp. 74–75). Accordingly, in addition to their traditional network of embassies, some states add permanent missions to international organizations in New York, Geneva, Rome or Vienna.
- There is also an increasing role played by **non-state actors** in the diplomatic game. This change affects multilateral diplomacy in particular. The "Earth Summit"[11] held in Rio in 1992 brought to light these new practices. Taking

---

[8]The use of the armed forces and economic pressures are the other two major instruments of foreign policy.

[9]The Vienna Convention on Diplomatic Relations was adopted on 18 April 1961 and entered into force on 24 April 1964. In 2015, 190 countries were signatory.

[10]There is, however, an embryonic diplomacy at the level of the European Union.

[11]The United Nations Conference on Environment and Development.

advantage of major international conferences, these new players express themselves, communicate, engage in lobbying, etc. The major types of private actors gaining ground in the international system range from non-governmental organizations on behalf of civil society to multinational companies representing the business community, etc. This system is no longer governed solely by states, even if they retain a prominent role.

– A third fundamental shift is the growing power of the **diplomacy of influence**, and the identification of *soft power* has led the way to this. Harvard University Professor Joseph Nye is credited with the now classic distinction between *hard power* and *soft power* (Nye 1990, 2004). In contrast to hard power, in which the state exercises its power of coercion by using traditional tools such as military power, soft power is the use of non-coercive means. As a third dimension of power alongside military power and economic power, soft power is "the ability to get what you want through attraction rather than coercion or payments" (Nye 2004, p. x). It relates to the larger field of the diplomacy of influence through which a country can "act as a decision maker and be perceived as such, with a few others" (Foucher 2013, p. 17).

– Finally, with the expansion of its **scope of intervention**, today's diplomacy shows a striking contrast with the diplomacy of the past, which limited its initiatives to political issues. A career in diplomacy could be then defined primarily by its own methods and the expected qualities of officers who exercised it (ability to negotiate, discretion, statesmanship...). Today these qualities clearly continue to be required of diplomats. But their missions have expanded over the twentieth century to the economy and other issues such as weapons of mass destruction or the environment, making the diplomat "an expert in negotiation, regardless of the area concerned" (Delcorde 2009, p. 4). The profusion of vocabulary stands testament: diplomacy today is at once *economic diplomacy* (which aims "to influence decisions about cross-border economic activities—export, import, investment, lending, aid and migration— pursued by governments and non state actors" Van Bergeijk 2009, p. 1), *energy diplomacy* (as practiced both by exporting countries seeking to derive the best political and economic advantage of their resources, and importing countries seeking to ensure security of supply), *nuclear diplomacy* (mainly dealing with non-proliferation issues), *environmental* or *green diplomacy* (driven by the rise of concerns about the protection of the environment), *climate diplomacy* (the most publicized environmental diplomacy component to date), *digital diplomacy* (relating to the use by states of the internet to expand their influence and protect their interests), *cultural diplomacy* (exercising influence through the promotion of literary and artistic heritage and of the national language) ... Science diplomacy has made a recent entry onto this long and open list.

Several concepts have been developed to account for these changes in the profession of diplomacy. Major UN forums have borrowed the term *stakeholder* from the business world to describe the proliferation of parties at major international conferences, which gave birth to the concept of *multistakeholder diplomacy*:

a "diplomatic innovative method aimed at facilitating the equitable participation of all parties concerned in discussions on and debates over particular issue or issues at stake" (Kurbalija and Katrandjiev 2006, p. vii). The term *mega-diplomacy* was also forged (Khanna 2011). A recent report prepared at the request of the Ministry of Foreign Affairs of Finland designed the term of *integrative diplomacy* (Hocking et al. 2012). All these formulations attempt to capture the changing conditions and emerging issues of diplomacy, now more transversal and inclusive than in the past.

What Relationship Between Science and Diplomacy? Stating the Problem
Can science issues influence international policy? Do diplomats really pay attention to what scientists say when negotiating the future of the planet? Is international scientific cooperation a factor of peace? Are researchers good ambassadors of their country? Is scientific prestige a particular form of cultural influence of a country? Does science diplomacy threaten the independence of the researcher? What is a scientific attaché for? For anyone wondering about the relationship between science and diplomacy, the questions keep coming and belong to various areas. Trying to answer them is the subject of this book. But how should we analyze this singular object of studies, how to clear this relatively unexplored field? To date, science diplomacy remains insufficiently studied. Although symposia and roundtables are dedicated to it and a quarterly journal, *Science & Diplomacy*, was recently launched, it is not yet a widely recognized area of academic study. No dissertation or summarizing volume has been devoted to it.[12] However, there are quite a few testimonials and comments, mostly from practitioners of science diplomacy in the United States. They provided valuable information for writing this book. More importantly, a study has paved the way: namely, the report co-authored in 2010 by the Royal Society and the American Association for the Advancement of Science. We will regularly refer to this work, to which we owe the first attempt of conceptualizing science diplomacy. Finally, we have enriched our reflection thanks to interviews with over 40 actors on the ground from a dozen countries.[13]

We also needed a set of questions to address the emerging reality of science diplomacy. It arose from two fields of knowledge, and science diplomacy is situated at the crossroads of the two. One is the field of *international relations*, due to the emphasis it puts on foreign policy: it provided us with keys for a better understanding of the soft power of science. Also, *science studies*, in which the social sciences, including sociology, question interactions between science and society: here we found benchmarks enabling us to elucidate the relationship between science and power, and the role of science in public decision.

We conclude these introductory remarks with two observations. First, and in many ways, the crossover between science and diplomacy is a variation of a more familiar theme, the meeting of science and political power. However, although it

---

[12]A recent exception is the collection of contributions co-edited by L. S. Davis and R. G. Patman in 2015.

[13]The list is given in the Acknowledgement.

can be seen as a projection in the international sphere of the relationship between science and politics, the relationship between science and diplomacy is not limited to it. The writings that deal with science and power on the international stage tend to focus on situations of war, in which the relation increases between those who hold power and those who hold some knowledge usable for military purposes. In the context of an armed conflict, the values of universality of science give way to the surge of patriotism, which may lead some scientists to put themselves more or less spontaneously at the service of opposing camps: the atomic issue during World War II has often served as an illustrative example. Dealing with science diplomacy instead means turning the spotlight onto situations and periods in history when diplomacy is the normal mode of communication between states, when the choice of dialogue and cooperation prevails over weapons. More diffuse and perhaps less spectacular than in a war context, the relationship between science and foreign policy during peace time gains in richness and complexity what it seems to lose in intensity.

Second, it was only in the new millennium that the words "science diplomacy" appeared. The interest taken in it is recent. Yet there were practices in the distant past that seem similar: the great voyages of exploration undertaken by European powers in the eighteenth century, although specific voyages had primarily scientific purposes (discovering distant lands, bringing back mineral or vegetable specimens, etc.), did not lack geopolitical goals. Closer to our time, the Cold War can be regarded as a period when interactions between science and foreign affairs asserted themselves, even if no one spoke of science diplomacy at that time. Thus, states have long considered science in their policy of influence. Far from ignoring it, they have practiced science diplomacy for decades, albeit not with the vocabulary that prevails today. It is only recently that science diplomacy has become a claimed approach, assumed by a growing number of countries as a component of their diplomacy of influence. In this sense, it is an emerging subject, to which little literature has been devoted to date. This book aims to fill this gap.

References

Davis, L.S., and R.G. Patman. 2015. *Science Diplomacy—New Day or False Dawn?* Singapore: World Scientific Publishing. 278 p.

Delcorde, R. 2005. *Les mots de la diplomatie*. Paris: L'Harmattan. 133 p.

———. 2009. *L'évolution du métier de diplomate*, vol. 10, *Annuaire Français de Relations Internationales*. Paris: Centre Thucydide—Analyse et recherche en relations internationales. 12 p. http://www.afri-ct.org

Foucher, M. 2013. Puissance et influence. Repère et référence. In *Atlas de l'influence française au XXIème siècle*, ed. M. Foucher, 10–17. Paris: Robert Laffont/Institut français.

Hocking, B., J. Melissen, S. Riordan, and P. Sharp. 2012. *Futures for Diplomacy—Integrative Diplomacy in the 21st Century*. The Hague: Netherland Institute of International Relations "Clingendael". 78 p.

Khanna, P. 2011. *How to Run the World*. New York: Random House. 272 p.

Kurbalija, J., and V. Katrandjiev. 2006. *Multistakeholder Diplomacy—Challenges and Opportunities*. Malta: Diplo Foundation. 204 p.

Merton, R.K. 1942/1973. The Normative Structure of Science. In *Robert K. Merton—The Sociology of Science: Theoretical and Empirical Investigations*, ed. N.W. Storer, 267–278. Chicago: University of Chicago Press.

National Science Board. 2012. *Science and Engineering Indicators 2012*. Arlington, VA: National Science Foundation. 592 p. http://www.nsf.gov/statistics/seind12/pdf/seind12.pdf

———. 2016. *Science and Engineering Indicators 2016*. Arlington, VA: National Science Foundation. 898 p. https://www.nsf.gov/statistics/2016/nsb20161/

Nye, J. 1990. *Bound to Lead: The Changing Nature of American Power*. New York: Basic Books. 307 p.

———. 2004. *Soft Power—The Means to Success in World Politics*. New York: Public Affairs. 191 p.

Royal Society and American Association for the Advancement of Science. 2010. *New Frontiers in Science Diplomacy: Navigating the Changing Balance of Power?* 32 p. http://diplomacy.aaas.org/files/New_Frontiers.pdf

Salomon, J.-J. 2006. *Les scientifiques entre pouvoir et savoir*. Paris: Albin Michel. 435 p.

UNESCO. 2015. *UNESCO Science Report—Toward 2030*. Paris: UNESCO Publishing. 820 p.

Van Bergeijk, P.A.G. 2009. *Economic Diplomacy and the Geography of International Trade*. Cheltenham: Edward Elgar. 240 p.

When we speak of science diplomacy, we use a vocabulary that did not exist prior to the present millennium. We also deal with an issue which has tended to be of recent interest these days, so we will try to specify the meaning being given to this concept. But behind the new vocabulary, science diplomacy appears to be rooted in the distant past: history gives evidence of ancient ties between science and foreign policy. This chapter, both conceptual and historical, is guided by this question: what is science diplomacy?

2.1 The Emergence of Science Diplomacy in Contemporary International Relations

2.1.1 In Search of a Definition

Being more and more used but not yet trivialized, the term "science diplomacy" comes from the Anglo-Saxon world, as do most of the relatively few writings relating to it. Norman Neureiter, a former science and technology adviser to the US Secretary of the Department of State, defines it as "an intentional effort to engage with other countries where the relationship is not good otherwise (Hsu 2011). The science allows you to deal with non-sensitive issues that both sides can work on together for the good for all". Vaughan Turekian, then-director of the American Association for the Advancement of Science (AAAS), expressed the same idea by stating that science diplomacy is "the use and application of science cooperation to help build bridges and enhance relationships between and amongst societies, with a particular interest in working in areas where there might not be other mechanisms for engagement at an official level".[1] Nina Fedoroff, who also served as science and

---

[1] As part of an interview reported in "Science as a tool for international diplomacy", http://cordis. europa.eu/news/rcn/30532_en.html. V. Turekian was formerly the Chief International Officer at

© Springer International Publishing AG 2017
P.-B. Ruffini, *Science and Diplomacy*, Science, Technology and Innovation Studies,
DOI 10.1007/978-3-319-55104-3_2

technology adviser to the Secretary of State and adviser to the administrator of the US Agency for International Development (USAID), stated that "science diplomacy is the use of scientific collaborations among nations to address the common problems facing twenty-first century humanity and to build constructive international partnerships" (Fedoroff 2009).

The use of science in diplomacy when other forms of dialogue are blocked, and placing it at the center of multilateral diplomacy to tackle global issues, are two major areas where science diplomacy manifests itself. This was already illustrated in the first pages of this book. But they do not cover all situations. A broader definition should be sought. The Royal Society-AAAS report provides the most convincing proposal. In this influential report, these two institutions have suggested to define science diplomacy from its three complementary components:

- Science in diplomacy

  Some areas of foreign policy need to be enlightened by science, which leads diplomats to seek input from the research community. The most obvious examples are found in international negotiations on global issues: scientific expertise and advice are essential to diplomats and policy-makers to address issues such as climate change, food security or energy. Science and scientific expertise are an aid to decision-making in foreign policy: to achieve its purposes, diplomacy must make effective use of science.

- Diplomacy for science

  Each country seeks to promote the national scientific community on the international stage and to facilitate cooperation with other countries: diplomatic and consular networks abroad are traditionally in charge of supporting the mobility of researchers (financial aid, visas) and assisting them in some negotiations (regarding intellectual property rights, for example). If it is decided to build major research infrastructures, this must been done by agreement of states through diplomatic dialogue, with shared costs and risks but also shared benefits through the participation of their researchers in multinational programs.

- Science for diplomacy

  When political tensions between countries do not allow for traditional diplomacy to express itself, scientific relations can be used to maintain or restore links. The role of science as a substitute for and vanguard of diplomacy is probably, among the three stated areas connecting science and diplomacy, the most original one, even if during some periods of international relations more so than others. US views on science diplomacy attach great importance to this aspect, which is well represented in the foreign policy of the United States.

---

the American Association for the Advancement of Science and the editor-in-chief of the quarterly journal *Science & Diplomacy*, launched in 2012. In September 2015, he was named the 5th Science and Technology Adviser to the Secretary of State.

Interest in science diplomacy, named as such, is recent. The reflections it raises and the policies it inspires mostly belong to the twenty-first century.[2] Several reasons explain this emergence of science diplomacy as a field of action and object of study. One of those reasons is the awareness of the existence of global issues, which has been widely shared since the second half of the twentieth century. We have already mentioned these challenges which need the input of science to be overcome: they relate to threats to the environment, human health or the safety of people. The international community is seeking solutions at a global scale in processes where scientific expertise and diplomatic negotiation intermingle. We will return to this in the last chapter of this book.

Another factor, more circumstantial, which has strengthened interest in science diplomacy, is the new stance of US foreign policy, a strong marker of which being the "Cairo speech" of 2011. In that speech, the President of the United States did not use the term *science diplomacy* even once, but he gave it substance by calling for a shift in relations between America and the Muslim world and, among others, by building upon on innovative partnerships in the fields of science, research and innovation to achieve this goal.[3] Science is therefore solicited, both for tackling global challenges and for re-launching international politics. A broader context, however, which highlighted science diplomacy, is today's acknowledged importance of *soft* power and of the diplomacy of influence.

2.1.2   Science Diplomacy Is a Form of Diplomacy of Influence

Bequeathed by the American political scientist Joseph Nye, the concept of soft power describes the "power of co-optation" by which a country can exert influence by playing the seduction and persuasion, its objective being to bring others to share its values, to reproduce its models, to "think like it". To achieve this, a country mobilizes resources such as its image, reputation, prestige, communication skills, attractiveness of culture, its science and technology profile, etc. The variety of modes of expression of soft power seems to have no limit: higher education with models provided by Harvard or the *Ecole Nationale d'Administration* (Lefebvre 2011), filmmaking (Hollywood), music (Korean pop), humanitarian action ("French doctors") . . . are a few examples.

We refer to soft power here because it paves the way for the analysis of the diplomacy of influence, which has profoundly changed diplomatic action. The concept of soft power is contemporaneous with the post-Cold War world and the

---

[2]Yet the combination of science to non-military objectives of foreign policy—in today's language, "Science for diplomacy"—has already been studied, notably with respect to the Cold War period (Doel and Wang 2002; Doel 1997).

[3]Another political voice was added in 2009 when the British Prime Minister Gordon Brown called on science to play a new role in international politics and diplomacy: "It is important that we create a new role for science in international policymaking and diplomacy . . . to place science at the heart of the progressive international agenda". This quote opens the *Royal Society-AAAS* report.

explosion of communication facilities of the late twentieth century and has been extended and enhanced since then. This is because the exercise of influence is not simply a matter of "soft power", and these two concepts are not synonymous. Today the term "smart power" is more commonly used as a concept which claims to be a synthesis of hard and soft power. This innovative concept in the field of foreign and security policy was defended in these terms by Secretary of State Hillary Clinton: "We must use what has been called "smart power", the full range of tools at our disposal—diplomatic, economic, military, political, legal, and cultural—picking the right tool, or combination of tools, for each situation".[4] The integrated approach to smart power then became the credo of the US administration, as evidenced by the official writings such as the President's *National Security Strategy* (The White House 2010) of the strategic plan for American diplomacy (Department of State–USAID 2010). This approach is also central to the foreign policy of emerging countries like Brazil, China and India (Foucher 2013, p. 10). This conceptual innovation is driven by observation of the international scene, in which the logic of confrontation does not disappear to the benefit of the logic of influence, but is integrated into a continuum which deeply renews diplomatic action.

In this light, science appears undoubtedly as one of the paths followed by the power of influence. Science can help a country to establish its reputation and gain a positive image on the international stage. In the midst of the Cold War, the successes of the Soviet Union in space exploration helped a lot to raise its image as a great science country, also able to achieve technological prowess. The Soviet Union drew an important benefit from this, particularly among non-aligned countries. In the post-war period, Japan built its image in the eyes of the world on its great capacity for industrial innovation. And in its process of aiming to move from being the world's factory toward becoming the world's laboratory, China validates the role that science can and should play in its progress towards the upper reaches of the global hierarchy of power (Lu 2010).

By its relationship with armament and defense policies (especially in nuclear and aerospace industries), science is historically linked to hard power. But here too, the center of gravity has shifted. Diplomacy includes the recognized importance of a knowledge society, thus taking note of "the growing interdependence between science and diplomacy in foreign policy of states" which founds what Philippe Lane called a "diplomacy of knowledge" (Lane 2016, p. 55).

Every element of the definition of science diplomacy given above links it, in one way or another, to the logic of influence, persuasion and exemplarity. Promoting one's interests and values: these ultimate goals of diplomacy find an ally in science. Science diplomacy is one of the forms of the diplomacy of influence. Highly performing research centers are attractive for foreign researchers, who in turn enrich them with their presence. The international reach of a country's research opens up the best opportunities to cooperate with others. It also facilitates the access

---

[4]This statement was made during her confirmation hearing before the Senate Foreign Relations Committee on 13 January, 2009.

of its researchers to international scientific expertise. These forms of influence through research activity and science will be detailed in the following chapters.

As we saw above, the diplomacy of influence takes a variety of forms. In the vicinity of science diplomacy, let's look at two of them in order to clarify what is special about our object of study. Let's first understand its position with respect to economic diplomacy. Research and technological innovation obviously create a link between science and business. But in the conduct of diplomacy, the scientific interests of a country and its economic interests can be distinguished by their time scale. Scientific issues are perceived as less immediate and less stringent than economic issues. The research performance of a country is not observed with the same acuteness as the success of its business in global markets. The balance of technological exchanges or the migratory balance of "brains" draw less attention than the monthly foreign trade statistics. Although investment in R&D is a driving force of economic competitiveness over the long term, successes and failures do not enter in the same way onto the scoreboards: in its relationship with diplomacy, science has a lot more autonomy than the economic sphere.

The positioning of science diplomacy relative to cultural diplomacy—another major form of diplomacy of influence—is more complex. Two approaches are possible. The first proceeds from a broad understanding of cultural diplomacy: it considers science diplomacy a subset of the latter. This view is widespread in the French administration, which generally includes university policy, science policy, the *Institut français* and *Alliances françaises* in the field of cultural diplomacy. As the former director of the *Institut français* Xavier Darcos stated, "the distinction between artistic culture and culture which embodies scientific genius is no longer operational" (Darcos 2011). In France, the support that the Ministry of Foreign Affairs grants to international scientific cooperation through its embassies falls within what is called the "cultural network". It is likely that this holistic view, which considers science as one of a number of vectors of a country's cultural policy and of its influence on the global scene, is justified by the ultimate purpose of influence. Science diplomacy is part of soft power, and so are the international promotion of national language or cinema, etc. Thus, culture and science can be combined in a single analysis (the external action of the state) all the more so when at the operational level they use the same kind of media and tools (embassies, cooperation budgets). Judging by international comparisons, however, it appears that other countries have a different approach, especially those which, as the United States and the United Kingdom do, have a clear awareness of their science diplomacy: they take it as a distinct type of cultural diplomacy and credit it with sufficient autonomy from cultural diplomacy such that it can, as we do in this book, be analyzed as such.

Finally, it should be stressed that, in addition to diplomacy in general being one of the attributes of sovereign states, science diplomacy is a responsibility of sovereign states and most often cannot be considered on a different level than their own. There is a French, an American or a Russian science diplomacy, but there is not as such a science diplomacy of the *Centre National de la Recherche Scientifique*, the National Science Foundation or the Russian Academy of Sciences.

This does not mean that these prestigious organizations should be excluded from the analysis, no more than any other autonomous institutions (universities, research institutes) which are the leading players in international scientific cooperation. But it is important to understand their proper place.

In the practice of science diplomacy, we distinguish two key levels. The first has to do with the visible interactions between issues of science and issues of diplomacy, which were identified with greatest clarity by the three components of science diplomacy identified by the *Royal Society-AAAS*. This is where the question of monitoring science diplomacy arises, and this will be discussed later. A second level is the practice of research actors on the global stage, when such practice contributes to expressing the soft power of science, for example through networking and image effects which benefit the country as a whole.[5] The second level is the one where the internationalization of research comes into the picture, in a diffuse but real manner, into the broad field of the diplomacy of influence.

But in order to avoid confusion, science diplomacy should not be confounded with the multifaceted field of internationalization of research. In particular, it should not be confused with international cooperation: as noted by Daryl Copeland, "by virtue of its direct relationship to government interests and objectives, science diplomacy differs from international scientific co-operation, which is sometimes commercially oriented and often without direct state participation" (Copeland 2011). The role of the state is to articulate in a subtle way the display of its international policy of research, as one of its state prerogatives, and the involvement in its implementation of researchers and scientific institutions, which are autonomous and operate through networks. As Nicolas Tenzer puts it in his analysis of the diplomacy of influence, "no central actor can dictate and federate all—unless at risk of causing adverse reactions, especially from private and academic actors—but everyone should have the feeling of contributing to what falls in the scope of a strategy of *national interest* although he or she also defends their own objectives" (Tenzer 2012, p. 8). It is the state that bears the ultimate responsibility of the national science diplomacy, without forgetting that, in the spirit of modern diplomacy, science diplomacy must be "integrative" in the sense that a large variety of actors must find their place in it.

To conclude this attempt of clarification, we propose to define science diplomacy as follows. At the intersection of science and foreign policy, a country's science diplomacy refers to all practices in which actions of researchers and of diplomats interact. These practices may be directly related to the interests of governments: this is the case when diplomats promote cooperation between scientists from different countries, whereas conversely international scientific relations facilitate the exercise of diplomacy or play an avant-garde role for it, and finally when scientific expertise helps governments and their diplomats to prepare and conduct international negotiations. Other practices of researchers find

---

[5]However, researchers could be associated informally with diplomatic efforts in the so-called "Track II Diplomacy". The Cold War period offers several examples.

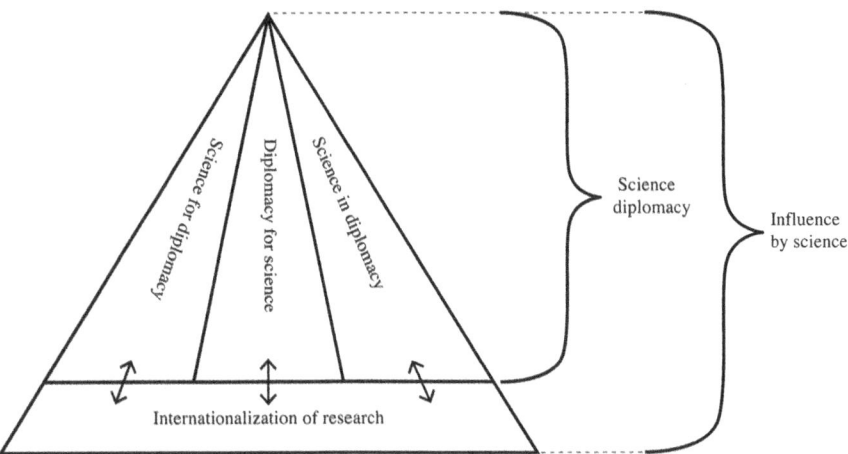

Fig. 2.1 Science diplomacy and internationalization of research in the diplomacy of influence

their place in a more diffuse manner in the broader context of a country's diplomacy, when they contribute to nourishing and strengthening its influence on the world stage, and ultimately to serve the national interest (Fig. 2.1).

2.2 A Historical Perspective

The vocabulary that prevails today should not obscure the fact that the involvement of science in foreign policy has a long history, as does the diplomatic role that has been assigned to it, sometimes without saying so explicitly. This articulation of science to diplomacy has followed several steps: if the great voyages of exploration appear as the prehistory of science diplomacy, the Cold War is the period when its foundations were laid.

2.2.1 Great Voyages of Exploration and Colonization

The discovery of America kicked off major voyages organized by European powers for the purpose of territorial and colonial expansion. They marked the conquest of the world by the Old Continent throughout the seventeenth and eighteenth centuries. At the beginning, scientific concerns were not in the foreground. The truly great scientific exploration trips were not organized until the advent of the Age of Enlightenment. This raises the question of how they fit in the international political and diplomatic game.

The first major scientific expeditions of the eighteenth century originated from Enlightenment philosophy and the opening that it allowed for the progress of knowledge. The great voyages of circumnavigation led the European crews to the

Pacific. One of these first trips was made by Louis-Antoine de Bougainville, a French naval officer who entered history as a navigator and explorer, but in his youth had been secretary to the French Embassy in London and member of the Royal Society, after publishing a treatise on integral calculus when he was 25. Naturalists, botanists and astronomers accompanied this trip around the world (1766–1769), during which they harvested a wide variety of observations.

During the same period, the Englishman James Cook made three long distance journeys. In the first of them aboard the HMB Endeavour (1768–1771), the Royal Society charged him with observing the transit of Venus from Tahiti and going in search of a "Southern Continent" (the future Antarctica) that was thought to exist. His second voyage (1772–1775) aimed at discovering this continent but did not succeed. But it attested to the newfound importance of science in geographical explorations: Cook took scholars with him representing the main branches of natural sciences. His third voyage (1776–1780—Cook died in 1779) aimed to discover the Northwest Passage by exploring the Bering Strait.

As opposed to the first great voyages of conquest of the world which Christopher Columbus initiated, intercontinental explorations of the eighteenth century were primarily motivated by scientific curiosity. The main disciplines which benefited from them were astronomy, natural history and cartography. But these trips, so decisive in the history of science, were not devoid of political agendas. Their backdrop was the struggle for the division of the world between the great European powers. They promoted the discovery and control of new shipping routes, the access to new areas or the strengthening of colonial ties. Monarchs were often sponsors of these expeditions. The journey of Bougainville was thus ordered by Louis XV. Taking Cook's voyages as models, Louis XVI sent Jean-François Galaup, Earl of La Perouse, on mission in the footsteps of the famous English navigator in order to complete his discoveries in the Pacific (1785, 1788). Empress Catherine II sent the *Slava Rossiy* expedition in search of the Northwest Passage (1785–1794) under Billings, a Brit who was Cook's astronomer during his third voyage.

In the large intercontinental ventures of the Enlightenment, purely scientific objectives and business objectives were intertwined, and it is difficult to separate them. Far from home, the scientist turned sometimes into a diplomat. This was the case during the journey to the southern lands (1800–1803) of Nicolas Baudin with his ships *Le Géographe* and *Le Naturaliste*, that brought back more than 100,000 specimens of plants and minerals to France. This immense contribution to the natural sciences and ethnology concealed a major political ambition: to gain a foothold in the South Seas before the British. Here is another example, on the side of the Russian Empire: the expeditions of vessels *Nadejda* and *Neva* (1803–1806) fed the Academy of Sciences in St. Petersburg with natural history specimens, but especially to establish a link between Russia and its territories in America and try to develop trade and diplomatic ties with Japan. Another example is that the laudable desire to know and map distant continents did not explain by itself the ability of the Bering Strait to attract several large expeditions from England, France and Russia, if one keeps in mind the geopolitical importance of this communication channel.

Special attention should be given to the French campaign in Egypt, launched by Bonaparte in 1798 which, with over 150 scientists, engineers, artists and technicians, "represented the culmination of the great politico-scientific expeditions" (Drouin 2003) (see Box 2.1).

Long journeys combining the interests of science and those of diplomacy continued at the beginning of nineteenth century. Such was the case of the Russian expeditions of the *Predpriyatiye* (1823–1826) and *Seniavine* (1826–1829), which brought back thousands of animal, plant and mineral specimens, while aiming at strengthening the presence of the Empire in Kamchatka and in the Alaska area.

The French expedition led by the ships *Le Rhône* and *La Durance* (1819–1821) recruited labor in Java and the Philippines for use in Guyana and also enabled them to study the acclimatization of plants brought from Asia on that South American territory. With the *La Favorite* (1829–1832) expedition, diplomatic and trade ties were established in India, Southeast Asia and Oceania, hydrological observations were made and natural history materials collected on a large scale.

Box 2.1 Bonaparte's Egypt campaign (1798–1801): An unprecedented combination of hard and soft power

The French campaign in Egypt refers to a military expedition led by Bonaparte and Kléber from 1798 to 1801 which aimed to create a new settlement in this country, paving the way to India, in a historical context marked by a strong rivalry with the British Empire. This campaign is a unique example of military conquest which was also pursuing scientific objectives. Bonaparte sailed from Toulon and took on board 36,000 soldiers and 2500 officers, but also more than 150 scientists, engineers, artists and technicians, organized in a "Commission of sciences and arts". Mathematicians Monge and Fourier, naturalist Geoffroy Saint Hilaire, chemist Berthollet and engineer Conté were part of the adventure: "Never, an army set out to conquer a country had taken in its wake such a number of scholars" (Beaucour 1970). The fleet landed at Alexandria and less than a month later, on August 22, 1798, the *Institut d'Egypte* was established in Cairo.

Replicating the *Institut de France* which was seen as a model, the *Institut d'Egypte* was organized into four sections (mathematics, physics and natural history, economics, literature and fine arts). It held 47 sessions up to1801. The scientists of the expedition presented papers, they organized field studies and produced a scientific journal (the *Décade égyptienne*). Their encyclopedic curiosity was deployed to observe the nature and resources of Egypt. Workshops, stores, and even a menagerie were created for the preservation of specimens. The Rosetta stone, through which Champollion was to succeed later in deciphering hieroglyphics, was discovered in 1799. All this amount of work eventually gave birth to the *Description de l'Egypte ou Recueil des*

(continued)

Box 2.1 (continued)

observations et recherches qui ont été faites en Egypte pendant l'expédition de l'armée française, a comprehensive 20-volume survey published starting in 1809 under the supervision of the Emperor.

Until 1801, when it ceased to exist, the *Institut d'Egypte* was a leading intellectual center. It was revived in 1859 in Alexandria under the name of *Institut égyptien*. Relocated to Cairo in 1880, it returned to its former name in 1918. Today it is under the supervision of the Egyptian government.

The Egyptian campaign was a military failure. But it demonstrated an unprecedented union in the colonial adventure between the power of armies and scientific curiosity. In addition to the technical support to the army in its conquest (including the study of digging a canal across the Isthmus of Suez), the work of the *Institut d'Egypte* made it possible to spread the Enlightenment in this part of the world and to promote a better knowledge of pharaonic Egypt in Europe. We should not talk here about science diplomacy, as the Egyptian campaign was a military expedition aimed at occupying foreign territories and submission by force of the people who lived there. But one cannot fail to be struck by the ambition of this multifaceted venture. The Egyptian campaign was conquest by force of arms but also aiming at civilization. It brought with it a breath from the century of Enlightenment. The prestige of science was part of the affirmation of grandeur, and this visionary approach has very modern accents. Being a military campaign and a scientific expedition at the same time, the initiative of General Bonaparte in Egypt demonstrated, in today's language, a unique combination of hard and soft power.

During the nineteenth century, the ascendance of major powers upon the world turned to imperialism. Colonization facilitated access to new territories attracting a scientific interest, and for some categories of scholars from dominant countries including naturalists and other scientists this enlarged the scope of what they could study: each colonial power encouraged the exploration of its colonies by "its" scholars. But, while the colonial power was supporting the work of scientists, the latter in turn sometimes supported the colonial policy. Thus, the French Academy of Sciences worked as a "relay of power" in a number of major scientific expeditions that it organized and supervised" (Blais 2004). This was particularly the case for expeditions to Algeria, which came under a broader political project, the colonial project. Written reports upon return from these trips and the discussions they led to showed the "symbiosis between academicians with the dominant ideology of the time (...), and showed again how the great voyage, as a science instrument, was put at the service of the state by the scientists themselves, and was a central element of the relationship between scientific institutions and power" (*ibid.*). Geographers played a particularly important role in colonization and foreign policy. In Morocco, "the bulk of the production of geographical knowledge was controlled and/or funded by the workings of the colonial administration, and directed by the needs of colonization" (Chouiki 2008, p. 116). Geographical societies sometimes favored overseas expansion, as noted by Jean-François Klein regarding the example of the Belgian Royal Geographical Society "whose grants to large African explorations hardly hid that they served as an excuse to Léopold II to prepare the annexation of the Belgian Congo and to have it accepted by the elite" or that of the Geographical Society of Tokyo, founded in 1876, 20 years before the Sino-Japanese War, which "revealed Japanese geopolitical intentions in the Far East" (Klein 2008, pp. 96–97). In today's language, we would rank these practices under the umbrella of "science for diplomacy".

2.2.2   The World Wars of the Twentieth Century

We cannot claim to deal with the relationship between science and diplomacy without commenting on the wars of the twentieth century and the roles that science and scientists have taken in them. War and diplomacy are only seemingly opposed in foreign policy, and the history of international relations can be seen as "alternating negotiations (Delcorde 2005, p. 78) before and after war".

War is the last resort of a country to defend its interests and its values, when diplomacy has failed. By its pressure on scientists—at least those whose skills can serve its purpose—war is a particularly interesting period for examining the relationship between science and political power, and in this sense it deserves attention here. Science entered the battlefield with the First World War, thus ruining the idealized image of its neutrality. We know the infamous case of Fritz Haber, laureate of the Nobel Prize for chemistry in 1918 for his work on the synthesis of ammonia, but who also invented poison gas, the first use of which he personally supervised in 1915 near Ieper.

The First World War demonstrated scientific and industrial resources through aircraft, submarines and chemical weapons. World War II demonstrated "an unprecedented investment of scholars in the war" (Pestre 2004, p. 16) with a "massive use of scientists and engineers in the preparation and monitoring of military operations"(*ibid.*, p. 14). The atomic issue vividly witnessed the anchoring of science considerations in defense and foreign policy. The events are relatively well-known. In August 1939, Albert Einstein signed a letter to the attention of President Roosevelt informing him of recent scientific advances for designing powerful bombs of a new type, and alerting him about the capabilities of Nazi Germany to carry out such a project. The acceleration of research that followed led the United States to engage in the construction of the atomic bomb: this was the Manhattan Project, conducted with the participation of the United Kingdom and Canada. An ideological competition soon added to the scientific and military competition between Germany and the Western camp led by the United States: in order to keep up, the USSR activated its scientists and engineers under the direction of physicist Igor Kurchatov, allowing the country to access the atomic bomb in 1949, 4 years after Hiroshima. In 1944, the Americans and their allies launched the

ALSOS operation, drawing its name from the unit in charge of monitoring the progress of Allied troops in Europe, capturing German scientists and appropriating all of what research Nazi Germany had produced not only in the nuclear field, but also in the fields of biological and chemical weapons. Not less than 1500 German researchers were brought to the United States and made available to the main winner of the war (via Operation Paperclip). The Soviets also gained resources and manpower from the defeated country through "Department 7" of their security services, some crumbs being left to the English and the French. Japan contributed in the same way to the booty of the winners.

World War II brought about a profound change in the recognition of science in the economy and in society. Joint working methods that soldiers, manufacturers and scientists had acquired during the conflict favored the use of mathematical modeling and simulation in planning for economic development. The "myth of science that won the last war" (Pestre 2004, p. 12) instituted an almost absolute belief in the ability of science and technology to solve all problems. World War II also had two particular effects with interesting international political consequences: the development of the "culture of secrecy" in order to protect national security, and the interest of intelligence services in a new area: the monitoring of the progress of science and technology among foreign powers. These new concerns would thrive on particularly fertile ground, that of the Cold War.

2.2.3 The Cold War: Science in the Ideological Competition

The very special relationship that developed in wartime between the scientific community and political and military power continued during the Cold War. But its expression became more complex and subtle. In the new setting of an ideological competition between East and West, the scientific community split between two attitudes: patriotism and internationalism, this second attitude experiencing a revival after being somewhat weakened during the World War.

Two analytical lenses should be favored in reading the history of the Cold War: first, the atomic question, and second, the place of science and technology in the competition between the blocs. In the wake of the Second World War, the atomic issue continued to be at central in questions relating to science and foreign policy. The escalation in nuclear weapons programs—and correspondingly, the issue of disarmament—were major issues of this period and are still very present. This was the time when, both in the United States and in the Soviet Union, physicists were mobilized to try to assess the progress of the other side in terms of nuclear weapons. The use of science for ideological purposes was not exceptional: for example, the "Atoms for Peace" proposal of President Eisenhower, while emphasizing cooperation and peaceful uses of nuclear energy, was part of the US strategy of containment of the influence of communism in the world.

The involvement of researchers in foreign policy was not limited to nuclear issues. It took place in the broader context of the ideological competition between the two camps. The US government employed scientists to thwart Soviet access to advanced technology that might be used for weapons. Other scientists worked with the government, sometimes in a clandestine manner, on matters relating to national security. The space sector emerged as a major place where scientific and technical issues and those of ideological confrontation intermingled. On October 4, 1957, the first successful launch of an artificial satellite sounded like a thunderclap. The USSR confirmed its superiority 4 years later by successfully completing the first manned space flight in the history. In only a few years, Sputnik and Gagarin had shaken the American supremacy in science and technology. These successes proved to be a powerful propaganda tool, giving credit to Khrushchev's watchword: "to rank first in the world according to both absolute level of production and output per capita". We know that the American response was not long to come. In 1961, President Kennedy set the objective: the human conquest of the moon by man before the end of the decade, and this was achieved 8 years later.

During the Cold War, however, the relationship between science and power was ambivalent. Another trend weighed against the patriotism of researchers, engineers and technicians of each of the camps who had offered their skills to the political and diplomatic objectives of their country. This other trend was internationalism of science, which expressed itself mainly in relation to issues of nuclear proliferation and disarmament. On July 9, 1955, Bertrand Russell and Albert Einstein appealed to the international scientific community to make it aware of the dangers of weapons of mass destruction. Mobilization translated into a first conference, held in July 1957 in Pugwash (Canada), which gave birth to the movement of the same name. This conference brought together scientists from East and West (including seven Americans, three Russians, three Japanese, two Britons and two Canadians, and one representative of each of Australia, Austria, China, France and Poland). Since then, nearly 300 Pugwash conferences have been held with a total of over 10,000 participants. They are recognized to have promoted dialogue on disarmament issues, especially in periods when official relations between the two blocs were difficult.

The creation of the United Nations Educational, Scientific and Cultural Organization (UNESCO) in 1945 can be seen as another manifestation of the internationalism of science. More generally, the creation and rise of the UN system has tended to move the center of gravity of the involvement of scientists into the multilateral framework. Many international treaties promoted this linkage between science and diplomacy: later, we will discuss the exemplary case of the Antarctic Treaty. The Partial Nuclear Test Ban Treaty, signed in Moscow in 1963, is another illustration.[6]

The Cold War was a period particularly conducive to the development of initiatives that appear today as founding moments in the history of science diplomacy.[7] Great progress in the international agenda of arms control and non-proliferation were possible thanks to the inclusion of the voice of science,

---

[6]This treaty covers the prohibition of nuclear weapons tests in the atmosphere, in outer space and under water.

[7]This is clearly evidenced by the historian Ronald E. Doel (1997).

"seen as a widely understood, non-ideological language that could be used to overcome, or at least mitigate international political differences" (Copeland 2011, p. 2).

2.2.4   After the Cold War

The fall of the Berlin Wall in 1989 and the dissolution of the Soviet Union in 1991 marked the entry into a new phase of history that for convenience we will call the "post-Cold War period". During this period, which was less predictable, more mobile and more open than the previous one, the relationships between science and foreign policy were given a new twist, which can be characterized in three ways.

Firstly, this is the period when pending issues of the Cold War were cleared up. A previously unaddressed question was raised: the reintegration into civilian activities of staff employed by the USSR and its allies in the production of chemical, biological and nuclear weapons. The future of these researchers, engineers and technicians of the military-industrial complex was a real concern for Western chancelleries, who feared that they could offer their talents to "hostile" countries such as North Korea, Iran or Libya. Following a US-Russian initiative, the answer came quickly in 1992 in the form of the creation, in Moscow, of a new intergovernmental organization, the International Science and Technology Center (ISTC),[8] with the primary mission "to give weapons scientists and engineers, particularly those who possess knowledge and skills related to weapons of mass destruction or missile delivery systems [...] to redirect their talents to peaceful activities".[9] The United States, Japan, Norway, the Republic of Korea and the European Union were the founding parties,[10] the beneficiary countries being Russia and six independent States of the former USSR. ISTC directs researchers and technicians holding know-how on sensitive matters to peaceful research activities, while giving them the opportunity to continue working in their country. Over the 1994–2014 period, more than 75,000 researchers and technicians of the seven beneficiary countries were funded by the programs of the Centre, whose headquarters have been in Astana (Kazakhstan) since 2015.[11]

---

[8] An equivalent center was created in parallel in Ukraine.

[9] Article 2 of the 27 November 1992 agreement establishing the International Science and Technology Center.

[10] As well as Canada, which later on withdrew from the organization.

[11] Dismantling arsenals was the other side of the clearance of the Cold War books: at the G8 Summit in Kananaskis (Canada) in June 2002, the Global Partnership against the Spread of Weapons and Materials of Mass Destruction (G8GP) was launched. Originally focused on countries of the former Soviet Union, in one decade it was able to dismantle more than 180 nuclear submarines, destroy thousands of tons of chemical weapons and secure thousands of radioactive sources.

Secondly, the end of the bipolarity of the world opened new spaces for science as well as diplomacy. The fall of the Berlin Wall dispelled the specter of nuclear war. International scientific cooperation, which each bloc presumably developed internally, was amplified with countries previously "on the other side". This forced the relaxation of state control over the mobility of researchers: "By 1990, international scientific exchanges had become so commonplace that the Department of State, which 30 years before had scrutinized each case, gave up trying to count them" (Doel and Wang 2002, p. 456). This was the period when the role of science in international relations raised a new interest, since this role was seen as one of the dimensions of soft power that arose in the post-Cold War era as a new horizon of diplomacy.

Finally, as we already noted, the last 20 or 30 years have seen the rise of concerns about damage to biodiversity, threats to the ozone layer and dangers of greenhouse gas emissions. Without any direct connection with the end of the Cold War, these issues have opened new definitions of the interface between science and politics. They have brought science into the heart of global geopolitics, as they themselves call for global solutions which are sought in the framework of multilateral diplomacy. Fully claiming their mission of whistleblowers, scientists were critical in these global issues entering into the diplomatic arena.

Contemporary with the rise of the diplomacy of influence, recently-named science diplomacy has emerged as one of its incarnations. But if we examine the history of international relations, there is no shortage of examples illustrating the relationship between issues of science and diplomacy. If looking at the great exploration voyages of the eighteenth century in terms of scientific diplomacy, one might determine that science diplomacy "boomed during the Cold War" (Copeland 2011, p. 2) since the paths of science and those of diplomacy often crossed in the ideological competition that prevailed. This scientific diplomacy was not referred to under that name, the vocabulary of the day having been different, yet the intention was there.

Finally, what allows us to draw a line between the interactions of science and diplomacy of the last 20 or 30 years and situations of the past is that in the present period these interactions stem from a conscious and increasingly displayed approach by states. Naming, claiming and conceptualizing science diplomacy is characteristic of the post-Cold War period. It is therefore important to analyze the challenges it poses to an increasing number of states which understand benefits they can draw from it in contemporary international relations.

References

Beaucour, F. 1970. L'Institut d'Egypte et ses travaux. *Souvenir Napoléonien* 255: 11–13. http://www.napoleon.org

Blais, H. 2004. Le rôle de l'Académie des sciences dans les voyages d'exploration au XIXe siècle. *La revue pour l'histoire du CNRS* 10. http://histoire-cnrs.revues.org/587

Chouiki, M. 2008. La géographie coloniale: engagement politique et flou identitaire—Le cas marocain. In *L'empire des géographes—Géographie, exploration et colonisation XIXe-XXe siècle*, ed. P. Singaravélou, 112–123. Paris: Belin.

Copeland, D. 2011. *Science Diplomacy: What's It All About?* CEPI-CIPS Policy Brief 13. http://cips.uottawa.ca/wp-content/uploads/2011/11/Copeland-Policy-Brief-Nov-11-5.pdf

Darcos, X. 2011. La diplomatie culturelle est un enjeu vital. *L'Express*, December 12. http://www.lexpress.fr/actualite/monde/xavier-darcos-la-diplomatie-culturelle-est-un-enjeu-vital_1060563.html

Delcorde, R. 2005. *Les mots de la diplomatie*. Paris: L'Harmattan. 133 p.

Department of State–USAID. 2010. *Leading Through Civilian Power—The First Quadrennial Diplomacy and Development Review*, 219 p.

Doel, R.E. 1997. Scientists as Policy Makers, Advisors and Intelligence Agents: Linking Contemporary Diplomatic History with the History of Contemporary Science. In *The Historiography of Contemporary Science and Technology*, ed. T. Söderqvist, 215–224. Reading: Harwood Academic Publishers.

Doel, R.E., and Z. Wang. 2002. Science and Technology. In *Encyclopedia of American Foreign Policy*, ed. A. De Conde, R. Dean Burns, and F. Logevall, 443–459. New York: Scribner.

Drouin, J.-M. 2003. *Les grands voyages scientifiques au siècle des Lumières*. Paris: Muséum National d'Histoire Naturelle—Centre Alexandre Koyré. 25 p.

Fedoroff, N. 2009, January 9. Science Diplomacy in the 21st Century. *Cell* 136(1): 9–11.

Foucher, M. 2013. Puissance et influence. Repère et référence. In *Atlas de l'influence française au XXIème siècle*, ed. M. Foucher, 10–17. Paris: Robert Laffont/Institut français.

Hsu, J. 2011. Backdoor Diplomacy: How U.S. Scientists Reach Out to Frenemies. *innovationnewsdaily*, April 8. http://www.innovationnewsdaily.com/196-science-diplomacy-soft-power.html

Klein, J.-F. 2008. La Société de géographie de Lyon: pour la Croix et la soie ? (1873–1900). In *L'empire des géographes—Géographie, exploration et colonisation XIXe-XXe siècle*, ed. P. Singaravélou, 91–109. Paris: Belin.

Lane, P. 2016. *Présence française dans le monde—L'action culturelle et scientifique*. Paris: La Documentation française. 128 p.

Lefebvre, M. 2011. L'ENA, outil du soft power à la française. In Enjeux et vecteurs de la diplomatie d'influence. *Mondes–Les Cahiers du Quai d'Orsay* 9: 45–51.

Lu, Y. 2010. *Science and Technology in China: A Roadmap to 2050—Strategic General Report of the Chinese Academy of Science*. Beijing and Heidelberg: Science Press and Springer. 138 p.

Pestre, D. 2004. Le nouvel univers des sciences et des techniques: une proposition générale. In *Les sciences pour la guerre: 1940–1960*, ed. A. Dahanand and D. Pestre, 11–47. Paris: Editions de l'Ecole des hautes études en sciences sociales.

Tenzer, N. 2012. L'influence des grandes puissances dans le monde et leurs stratégies pour l'avenir. *Mondes–Les Cahiers du Quai d'Orsay* 9: 7–14.

The White House. 2010. *National Security Strategy*, 52 p. http://www.whitehouse.gov/sites/default/files/rss__viewer/national_security_strategy.pdf

Science Diplomacy as a National Issue

In all its forms of expression, diplomacy is based on the sovereignty of states. This framework is the starting point of our analysis: science diplomacy is a national issue. We examine its mechanisms here. But in order to understand the relationships between the world of scholars and the world of diplomats, it is first necessary to observe the characteristics specific to each of them.

3.1   Universal Science and the Power of Nations

Science diplomacy involves two very different worlds. The world of science, i.e. the production of knowledge, is, on the one hand, represented by the iconic character of the scientist, or researcher. And, on the other hand, there is the world of foreign policy, which features the character of the diplomat. In two essays which have remained highly acclaimed,[1] a century ago Max Weber contrasted the "scientist" and the "politician". Should we today contrast the figures of the researcher and of the diplomat with the same vigor?

3.1.1   The Scientist and the Diplomat: Contours of Two Types

"Scientists and diplomats are not obvious bedfellows" (Royal Society-AAAS 2010). With a touch of British humor, this statement sets the starting positions: science diplomacy features two characters—the scientist and the diplomat—driven by values and interests so different that it seems difficult to imagine that they have something to share.

---

[1]"Science as a Vocation" and "Politics as a Vocation" are two famous lectures that Max Weber gave in Munich in 1917 and 1919 respectively.

© Springer International Publishing AG 2017   27
P.-B. Ruffini, *Science and Diplomacy*, Science, Technology and Innovation Studies,
DOI 10.1007/978-3-319-55104-3_3

Sociology has addressed the first of these two figures, the scientist, in line with the original writings of Robert Merton. It has highlighted the core values of science: disinterest, rationality, transparency and universality. But while there is a sociology of science, there does not exist a sociology of diplomacy per se, which singles out social characteristics and conditions in which the profession is exercised that could be easily compared to those of scientific activity.[2] Here, we will explore two dividing lines.

3.1.1.1 Science Without Borders Versus National Political Spaces

"There is no national science just as there is no national multiplication table. What is national is not science". In just two sentences, the Russian author of short stories and playwright Anton Chekhov seems to have closed the debate. In its essence, science knows no borders. Studying the structure of matter or living organisms is to try to unlock the secrets of nature and thus involves asking questions of interest to mankind as a whole. Without overlooking the influence that the original environment or the era may have on the way scientists conduct their research, the outcomes transcend national loyalties and are universal in scope. And when the object of study cannot be detached from its historical and cultural context, as is almost always the case in social sciences, the scientific method, based on the rationality of discourse and verification of assumptions in ways that minimize the subjectivity due to the observer, leads to results that are not national objects, but—like the outcomes of natural science—additions to the common heritage of knowledge.

The language of science and its methods keep it away from national frameworks. In each discipline, the community of researchers transcends national borders. Methods and results are compared on the world stage, during conferences where the shift of the frontier of knowledge is measured at regular intervals. Validation of results is gained through "peer reviews" made by colleagues who, without distinction of nationality, conduct critical and anonymous review of research articles for publication in professional journals. Thus, both by its purpose and its methods, science is borderless.

Comparing the researcher to the diplomat in this respect is almost trivial. The latter, who orchestrates the mechanics of political relations between countries on the ground, derives legitimacy from the existence of nations. If science knows no boundaries, diplomacy is an art that results from the division of the global space into sovereign nations. The plurality of nation-states is essential to the diplomat, just as borders are to customs officers or national currencies to foreign exchange traders. The diplomat is an active agent of dialogue between different national communities. But there are agendas in such dialogues not all of which clear: the

---

[2]This does not mean that social scientists are not interested in the world of diplomacy: see for example Piotet et al. (2013), Bazouni (2005), Delcorde (2005).

neutrality of the researcher is as opposed to the bias of the diplomat, who by construction serves the interests of his or her country.

3.1.1.2 Search for Truth, Fairness and Transparency Versus Strategy, Manipulation and Secret

The scientist is driven by an ideal, which is to advance knowledge in an impartial and disinterested way. These noble values are associated with practices specific to the scientific community. Any new knowledge is accessible to everybody: research must be transparent, its results are intended to be published; they can be checked by contradictory validation procedures and possibly challenged. Science is an open field. In the language of economists, science is a "public good": anyone can have free access to its achievements and can use them without reducing the possibility for others to do the same. The circulation of people is added to the flow of information, as a necessary condition for the facing off and sharing of ideas, and therefore for scientific creation.

The diplomat is governed by other principles. As a servant of state power in the sphere of international relations, the diplomat "must use his ears, not his mouth".[3] It has been described as "an honest man sent to lie abroad for the good of his country"[4] and when he/she negotiates, which is one of his major tasks, they are said to not exclude trickery or concealment. These common views on the diplomatic profession contain strong prejudice and caricature elements. As Delcorde writes, "the diplomat who lies rarely has his way (. . .). A reliable diplomacy is in the long run preferable to Florentine intricacies with no holds barred" (Delcorde 2005, p. 96 and 102). But they give a measure of what can distinguish between the diplomat and the researcher. For the latter, the search for truth and the impartiality of the method are core values. To the diplomat, they are of relative importance and do not always find their place in the use of negotiation techniques such "trial balloons", threats, bluffing or constructive ambiguity (Loriol et al. 2008, p. 82). Influence, persuasion, balance of power or strategic reasoning are key words of the diplomat's repository. They are not part of the researcher's one. And the interests of science and diplomacy can even be opposed. A secret, which "is the very soul of diplomacy"[5] is contrary to the scientific ethos as defined by Merton. In extreme cases, the values of science are denied when knowledge is manipulated for purposes of domination or when science is used as an instrument by a totalitarian ideology.

Ultimately, the scientist and the diplomat would, in their canonical missions, fall within two disjointed and even contradictory worlds. A "horizontal" world on one side, that of knowledge, in which researchers answer questions asked by their discipline, and in which results are evaluated in a collegial way, a world full of

---

[3]This quote is attributed to Japanese diplomat Komura Jutaro.

[4]This judgment of Sir Henry Wotton, an English seventeenth century diplomat, is quoted in the report *Royal Society–AAAS*.

[5]F. de Callières (1716), *De la manière de négocier avec le souverain*, quoted by R. Delcorde (2005), p.105.

people driven by unselfish curiosity and who practice transparency and impartiality. And a "vertical world" on the other side, an action-oriented world, made of power, hierarchy and command, and embodied in the sphere of international relations by the diplomat who is the promoter of the interests of his country, possibly detrimental to the interests of others, and who pursues his goals using the weapons of negotiation but also of manipulation or secret. The scientist and the diplomat belong to very different cultures, which Daryl Copeland sums up as follows: "When diplomats or politicians talk about international policy, you rarely hear anything about science and technology. Similarly, when scientists get together to discuss their work, it is rarely in the context of diplomacy or international policy. Indeed, scientists, besides being notoriously poor communicators, tend to cherish their independence from politics and government. The skill sets, activity time frames and orientations of the two groups differ markedly" (Copeland 2011).[6]

Yet, although they are motivated by goals and values which seem irreconcilable at first glance, the world of science and the world of diplomacy do not ignore each other. Better still, these worlds interact, and without such a dialogue science diplomacy simply could not exist. And, while we start from a point where there is strong contrast and well-entrenched positions prevail, these differences are becoming smaller in the momentum of action.

3.1.2   The Scientist and the Diplomat: A Proximity of Convenience?

This confused scene that we have just set is too simple. There is not on one side a world without sin—a paradise of researchers with noble qualities—and on the other side an earthly life with its ups and downs, its arrangements and compromises, which would be reserved for diplomats. It is time for the researcher's dreams to meet reality, and it is because he lives on earth that he crosses the path of the diplomat. Looking beyond assumptions and prejudices, the scientist and the diplomat are closer than they appear for three reasons.

First, the more science becomes embedded in industrial processes, incorporated into innovation, expressing itself in technical progress, etc., the more it enters into the sphere of attraction of politics. With the advent of modern science in the nineteenth century, theoretical knowledge became difficult to separate from its applications. This is the case for what is commonly called applied research. But

---

[6]A recent evaluation confirms this view: "Many of the misunderstandings between both diplomatic and scientific worlds come from reciprocal fears that must be overcome: the first one does not understand the reluctance of major research organizations or institutions to accept the implementation of partnerships intended primarily to bring France and foreign countries closer together, with the hope of diplomatic benefits; the second one fears the dilution in short term objectives of scientific excellence, the only basis for lasting cooperation in the scientist's eyes", Ministère de l'enseignement supérieur et de la recherche and Ministère des affaires étrangères (2014), *La coordination de l'action internationale en matière d'enseignement supérieur et de recherche*, p. 83.

this also applies to basic research: let's follow here Jean-Jacques Salomon, for whom "there is today only a small part of what we call "basic research" that is not associated in one way or another to projects related to defense or industry" (Salomon 2006, p. 13). This development "put researchers under the growing dependence of states (...). The rise of modern science as a collective praxis puts the dual loyalty of scientists towards science and towards humanity under the common law of national loyalties" (Salomon 1970, p. 322 and 318). As the processes of science are performed, scientists are confronted with the reality of national rivalries: the entry into the atomic age and the Manhattan Project provided the strongest illustrations. Because science does not exist in a weightlessness state above society, but is intended to become one with society in order to promote its progress through its applications, science enters the field of power relationships, which are orchestrated on the international scene by foreign policies of states. Therefore, "interstate relationships involve science and address science; they include it so well in diplomacy or war that there exist researchers to live them and symbolize them, in the very exercise of their duties, as ambassadors and soldiers" (Salomon 1970, p. 324).[7]

Second, the methods of organization and funding of science are national and related to policies carried out in a national framework. There are several reasons for this, which brings us back to the fact that science is of a public good nature. Advances in basic research depend on long-term investments with uncertain results, and this limits their funding to public sources. This public funding is irreplaceable for basic research and remains the basis on which the entire research and innovation system is grounded. There is no significant innovation without a strong basis of science, as evidenced for example by available data on the relationship between science and patents. Innovative firms are dependent on investments, most of which are public: infrastructure and networks enabling innovation, human capital formed by the higher education system and other intangible factors, such as access to data. Public investment provides the impetus for innovation, as shown by the examples of internet browsers or the human genome project (OECD 2010).

The importance of this research being framed in a national context underlies recent analysis of "national research and innovation systems". As written by Bruno Amable, "the main assumption is that national structural differences play a role in national modes of innovation, competitiveness, sector specialization and possibly growth" (Amable 2003, p. 372). Based on the fact that patterns of knowledge accumulation are very different across countries, they are classified "according to the structures of their scientific and technical systems and the manner in which science and technology interact with other areas of the economy" (*ibid.*).

The influence of public authority has led to speaking of the "nationalization" of science, in the words of David Edgerton (1997) as recalled by Dominique Pestre

---

[7]J.-J. Salomon notes then that "the specificity of the scientist, whose business, culture, values were indifferent, if not disobedient to the political sphere, vanishes into specific behaviors by which he expresses and represents in turn the state in its relations with other states".

(Pestre 2004, p. 18): in the second half of the twentieth century in particular, science has become a "means of action available to the state to strengthen the nation: it is built away from the state and its laws, it is built away from major national (in the US) or nationalized industrial companies (in France)" (*ibid.*, p. 19).

Finally, it is important to distinguish between science and the scientist as a person, the man or woman who embodies it, with his or her greatness of spirit, but also possibly his or her weaknesses. Self-denial is one of the values of science. But just as he is not stateless, the researcher is not necessarily selfless. Hunger for power or money can compromise the ethics of science. Competition for funding, the iron law of "publish or perish", the unbridled pursuit of fame, can transform healthy emulation into a fierce competition that does not exclude low blows. This may question the impartiality and the integrity of the researcher, as several cases of spectacular scientific fraud have shown.

In the search for points of convergence between the scientist and the diplomat, who is a national character par excellence, there is a useful observation: although upheld by universalist values, science does not escape some form of patriotism and it is difficult for the researcher to avoid any national preference. The 1930s and 1940s have offered the most striking illustration of the mobilization of researchers at the service of a national cause. But it is not only through a conflict between nations, whether an open war or a cold war, that the nationhood of the researcher can express itself.

From a cognitive point of view, there is no American, French or Russian science and Chekhov is right, nothing is national in the multiplication table. But there is an American, French or Russian way of "doing science". Science is universal, but the ways by which knowledge is produced remain largely national, as already noted. Let us introduce here a new dimension, that of the influence of national history, culture or education on the behavior of researchers. This influence is recognized with the existence of "national schools" such as the Russian school of mathematics, the German school of chemistry or the French school of sociology, which are part of the legacy of the history of science in these countries and continue as important centers of accumulation of intellectual resources.

A strong example of the influence of national belongingness in scientific production is offered by one of the best representatives of the French school of mathematics, and deals with the issue of language: "Mathematics is virtually the only science where, in France, researchers commonly continue to publish their work in our language. It is usually argued that it is because the French mathematical school occupies such an exceptionally strong position in the world that it can maintain such use. I am convinced that the cause and effect relationship is reversed: it is in so far as the French mathematical school remains committed to the French language that it retains its originality and strength. Conversely, the weaknesses of France in some scientific disciplines may be related to linguistic abandonment. The springs of this causality do not belong to the scientific order, but to the human one; they relate to the psychological, moral, cultural and spiritual conditions that make possible scientific creativity" (Lafforgue 2005).[8] By its power to structure thinking, language offers the researcher unusual bases for reflection that enriches his/her creative potential. Laurent Lafforgue invites others to share his belief with these words: "Scientific creativity is rooted in culture, in all its dimensions—linguistic and literary, philosophical, and even religious".

Because it cannot be totally detached from the national framework, the production of science may support the policy of influence of states: a country can attract by its way of teaching and its university model—it can attract and spread its influence by its way of doing science. The British or French style in scientific research is not the American style: as Charles Halary writes, "in these two countries, the concept of basic research is culturally usually accepted as distinct from that of technological development. In France and Britain, the aristocratic ideal of science as a game of the mind is still alive today" (Halary 1994, p. 51). Therefore, the practice of science feeds on the "soul of peoples", which was dear to André Siegfried (1950). In international relations, it is a way to export part of one's national identity: it is a vector of soft power.

Being an asset in influence strategies, the national aspect of scientific production creates a bridge between the researcher, a universalist by essence, and the diplomat, who is necessarily a national character. But how far can the approximation, and possibly the convergence, of interests of actors of these two worlds go? In its concrete achievements, science diplomacy brings mixed answers, which the following developments will illustrate on many examples.

3.2 The Challenges of Science Diplomacy

Science and technology—and more broadly, knowledge—are today fully recognized as essential resources for economic and social development. This is what historical observation confirms: social progress and material well-being of peoples depend in the long-term on efforts in education and R&D. It is not surprising, therefore, that support for research and technology is one of the priorities set by governments. But scientific and technological development does not bring only domestic benefits: it is also an asset to any country wishing to hold its position in the world and to assert its interests. These concerns express themselves in different ways in the sphere of foreign affairs and diplomacy. We summarize them in three key words: attraction, cooperation, influence.

3.2.1 Attraction

"The only wealth and genuine force is people". The famous aphorism of Jean Bodin[9] applies perfectly to scientific research. The human factor is at the heart of

---

[8]Laurent Lafforgue was a Fields Medal winner in 2002.

[9]Jean Bodin (1529–1596) is a French philosopher, economist and political theorist, author of the *Six livres de la République* (1576) from which the famous maxim is derived.

the production of knowledge. Crucial as they are, technical equipments and information resources that must be mobilized are nothing without the intelligence of the researcher: research activity is a "grey matter industry", and this production factor has several outstanding features. First, it yields its fruit only after a long upfront investment: 8–10 years of post-secondary study are needed to get a doctoral degree, which is the minimum background required to enter into a career in research. Second, it has a high maintenance cost: the researcher must constantly absorb all the new information relating to his field in order to take into account the results of research done by others in his own work. Another feature is that human capital devoted to research is almost always mobilized within teams, which makes research activity a collective co-production involving researchers, research engineers and laboratory technicians. Finally, the grey matter is internationally mobile, and hence there are policies to increase the attractiveness of a country to researchers.

National research systems are directly competing in order to take possession of "brains" of researchers. They try to influence the global sharing of grey matter in order to be more competitive in the production of new knowledge. Accumulating science-intensive human capital which elevates the national endowment by importing brainpower is the goal. Attracting the best foreign students is similarly motivated, especially students in doctoral studies.

Researchers are an important component of all highly skilled workers who emigrate, next to other categories of talents such as engineers or physicians. Brain drain is the traditional approach to the international mobility of researchers. The optics of brain drain, which draws attention to the loss of resources faced by countries of emigration, is to be compared to that of brain gain, which emphasizes the benefits accruing to host countries. But next to brain drain and brain gain, which describe the two sides of lasting or permanent expatriation of researchers, there is a broader approach, that of brain circulation, which attempts to encompass all forms of international scientific mobility: exodus and return of grey matter, but also temporary mobility (from a few days to several months) of researchers. Temporary mobility, that is to say, the to-and-from of researchers between their laboratory and those of their colleagues abroad, is part of a cooperation-based approach. By contrast, national competing strategies are at work when it comes to attracting and retaining grey matter.

Global brain circulation is difficult to capture statistically. Data collected by international organizations (OECD and UNESCO) focus on highly skilled workers, of which researchers are a part. There is no overall picture specifically focusing on researchers, both in terms of stock (census of researchers settled abroad on a long-term basis) or flow (mobility of researchers from one country to another). To illustrate the variety of national situations, we refer here to some results of the 2011 *Globsci* survey which addressed 17,182 researchers from 16 countries in order to study the different models of international mobility. Researchers were identified from the publications to which they had contributed in 2009 in four disciplines: biology, chemistry, earth and environmental sciences, and materials (Franzoni et al.

2012).[10] The study provides valuable information on the reasons that led researchers to leave their country, and analyzes the influence of mobility on the quality of international research networks. Here we retain information that assesses the importance of the "scientific diaspora" (the number of researchers who arrived in their new country of residence after the age of 18). By destination, Switzerland is the country with the highest proportion of researchers from abroad (56.7%), followed by Canada (46.5%), Australia (44.5%), the US (38.4%) and Sweden (37.6%). At the other extreme, the lowest percentages of researchers from abroad are found in India (0.8%), Italy (3.0%), Japan (5.0%), Brazil (7.1%) and Spain (7.3%). France occupies an intermediate position (17.3%). If one looks at the country of origin of the diaspora, India tops with a percentage of 39.8% of its researchers abroad, followed by Switzerland (33.1%), the Netherlands (26.4%) and the UK (25.1%). Conversely, the share of expatriate researchers is very low for Japan (3.1%), the US (5.0%), Brazil (8.3%) and Spain (8.4%); the corresponding figure for France is 13.2%. Finally, the United States is the top destination for migrant researchers from almost all the countries under review. Although the survey covers only a single year and four research disciplines, and is limited to 16 countries, the survey has the advantage of allowing international comparisons: in a quick shortcut, it shows that Swiss and Indian researchers are more mobile, and Japanese and US researchers less mobile.

To illustrate how crucial it is for a country to successfully attract brainpower, one need only watch the gap between the high performance of the United States in research and the poorer outcomes in terms of training (with only 24.5% of the US population accessing tertiary degree against 30% in France, Germany or Japan) (UNESCO 2010, p. 19). What would research in the US be like without the contribution of researchers originating from elsewhere? In the US, one in three postgraduate degree holders is not American. The proportion of researchers from abroad, and not just in the four fields of study included in the survey mentioned above, is increasing: they accounted for one-quarter in 1970 and one-half in 2010 (National Science Board 2012).

Let us now turn to the case of students. Attracting foreign students and training them at home is a goal shared by all countries that aspire to be influential. The reasons are easy to understand because the benefits over the long term far outweigh the immediate costs. After graduating, should the foreign student decide to stay and work in the country that welcomed him, he will contribute to its development through his qualification and skills. And if he decides to return to his country of origin, he usually behaves like an ambassador of the country that trained him.[11]

---

[10]The 16 countries are: Australia, Belgium, Brazil, Canada, Denmark, France, India, Italy, Japan, Netherlands, Spain, Sweden, Switzerland, United Kingdom and USA. These 16 countries represent 70% of all articles published in these disciplines. Reviews having made it possible to identify researchers were selected at random in all scientific journals indexed by the *Institute for Scientific Information*. Due to difficulties in conducting the survey, China has been left out, which is a limitation of this work.

[11]By maintaining links between graduates and the university or school that trained them, alumni associations participate in the diplomacy of influence of a country.

Student mobility is beyond the scope of our study,[12] except that of doctoral students. In the competition among nations in attracting or retaining grey matter, doctoral studies are of strategic importance: welcoming students from abroad for a PhD means expanding the pool that the host country can draw from to ensure the renewal of its research staff. The propensity of PhD holders to stay in the country that welcomed them is high indeed, whether as post-doctoral fellows or permanent research fellows (Auriol 2010). Sixty-five percent of foreign students who obtained their doctorate degree in the United States in 2001 were still in the country 10 years later (National Science Board 2014).

This phenomenon is not limited to the United States: the labor market for PhDs is more internationalized than for holders of any other degree. According to a 2011 OECD study, mobility has affected 14% of doctorate holders during the first decade of this millennium, with the United States remaining the top destination, but intra-European mobility is growing, particularly to the benefit of France, Germany and the UK (OECD 2011). These figures exemplify the political importance attached to the "production" of PhD holders and the attraction of foreign doctoral students.

Various means are implemented to support the attraction of grey matter, whether professional researchers or doctoral students. These include accommodating policies of granting visas, scholarships or material support to set up on the national territory, assistance for the return of expatriate researchers, etc. These actions mobilize a variety of players, some of which implement the government's policy (such as ministries) and some others (such as universities) implementing their own policy: some scholarships for foreign students, for example, may be covered by national public programs but others also originate from universities, research centers, local authorities or companies.

Diplomatic apparatuses play a key role in promoting the attractiveness of national research systems. Embassies disseminate information on scholarship programs or notices of recruitment of researchers by the country they represent. They also act through their scientific counselors and attachés to facilitate contacts with the scientific diaspora, and encourage expatriate postdoctoral students and researchers to return home. Consulates abroad relay national immigration policies abroad and are an essential link in the granting of visas for research purposes. Finally, ministries of foreign affairs may decide to devote part of their budget to the attraction of grey matter; but this form of involvement—particularly relevant for the case of France—is an exception rather than a rule among countries.

---

[12]According to UNESCO statistics, the number of international students in 2002 was 2.1 million. It amounted to 4.1 million in 2013. Regarding students attractiveness, English speaking countries hold in global competition a clear advantage: among them, the United States, the United Kingdom, Australia and Canada are the top four destinations, gathering in 2009 40% of students international mobility, although their share of world total is declining (it was around 50% in 2002). See R. Choudaha and C. Li (2012).

3.2.2 Cooperation

International cooperation, whether bilateral or multilateral, is most consistent with the values of sharing and universality of science. Countries pool their material and intellectual resources in order to obtain scientific results that not only meet the goals they have set, but also bring benefits to all others and enrich the world's knowledge heritage. Unlike the approach of attracting scientific capital—where all captured/attracted grey matter represents a loss for competing countries—cooperation is mutually advantageous. The logic of cooperation replaces the logic of competition described above: in official speeches, there is no country that does not claim its willingness to cooperate.

International scientific cooperation can be grasped and concretely measured by the number of co-publications which it generates. An international co-publication is written by at least two authors belonging to research institutions from at least two different countries. From 1988 to 2009, the share of co-publications increased from 8 to 23% of all research papers (National Science Board 2012). This growing internationalization of scientific production is proportionate to the deepening of international cooperation that makes it possible.

Citations are another indicator which confirms this trend. When frequently cited by other researchers, a research article provides evidence of its impact and usefulness in the collective process of knowledge production. In most countries, citations of articles from the international literature have increased at the expense of citations of purely domestic papers. In the US, the European Union and China, half of all citations refer to articles that include at least one author from another country (National Science Board 2012).

By the values of dialogue and sharing it embodies, international cooperation regarding research is an important area for diplomatic action. There exists a diplomatic accompaniment of cooperation, which feeds the "diplomacy for science" pane in the three reference perspectives in relations between science and diplomacy. The political will of countries to cooperate in the field of research is reflected in the signing of bilateral scientific and technological agreements. Ministries of foreign affairs of both countries are stakeholders of such governmental agreements. Ministries of research are associated with it, sometimes formally, being among the signing parties of the agreements, and always functionally, given their role in the implementation of research policy. Such agreements are generally concluded for periods ranging from 3 to 5 years.

Scientific and technological bilateral agreements express the joint will of the countries signing them to deepen their exchanges and joint actions. Authorities often display the number of such agreements as testimony to their openness to cooperation. China has thus punctuated its entrance onto the world stage of science by signing such agreements. In 2010, there were 30 protocols signed between that country and the United States, all covered by the framework agreement signed between them in 1979, when diplomatic relations were established.

Intergovernmental agreements, which merely state very general objectives, are implemented in practice through periodic bilateral meetings.[13] It is the role of "joint commissions" to oversee implementation and to identify specific areas where both countries wish to see their researchers cooperate. Depending upon agendas, representatives of research institutes or universities participate in these commissions, alongside representatives of ministries of both countries. For example, it is a joint commission convened every 2 years that ensures the proper implementation of the cooperation agreement between the United States and China; this commission is co-chaired by the Chinese Minister for Science and Technology and the Science Advisor to the US President.[14]

The role of these bodies is particularly important when they have the authority to allocate funds to the achievement of higher priority cooperation. But financial support which diplomacy may grant to scientific cooperation does not necessarily lend it to the channel of joint commissions. In some countries—France for instance—the Ministry of Foreign Affairs dedicates resources to support bilateral scientific cooperation: the department subsidizes nearly 100 scientific cooperation and research programs (Ministry of Foreign Affairs 2013, p. 5).

3.2.3 Influence

Among the forces which drive science diplomacy, the desire to influence is probably the one that is more diffuse. Influence is the capacity of a country to weigh in on decisions and events. Influence manifests itself especially through the attractiveness of the national research system or the ability to cooperate. It cannot be reduced to the projection of a positive image, as conveyed by a successful and recognized scientific community. It is important to go further by identifying the real levers of influence. We distinguish three: presence in international organizations, expertise, and hosting scientific organizations and major research infrastructures in the country.

3.2.3.1 Presence in International Organizations

International organizations include a wide range of over 350 entities, the largest share of which falls within the United Nations system, but also includes European institutions, the so-called "coordinated organizations" such as the OECD and NATO, and other international organizations such as the European Molecular Biology Laboratory (based in Heidelberg) and the European Southern Observatory (ESO).

Nearly 190,000 people were employed in international organizations in 2011, with about half of them in the United Nations system. These figures cover several

---

[13]For instance, the United Kingdom regularly organizes bilateral summits on science and technology issues with Brazil, China, India, Russia, South Africa and South Korea.

[14]The joint commission and an executive secretariat oversee the implementation of the framework agreement. An executive secretaries' meeting takes place in years when the joint commission has not met. It takes stock of cooperation activities covered by the agreement and proposes changes and other amendments to the joint commission (Suttmeier 2010, p. 28).

categories of employees: international civil servants, recruited competitively, and with career prospects in these organizations; officers seconded or assigned by their national administration for shorter or longer periods; and staff recruited by contract by the organization itself in order to ensure execution of permanent tasks.

The available statistics do not specify "scientists" within the total workforce of international organizations. Some international organizations have missions which clearly mobilize fields of knowledge and expertise in the sciences, such as the UNESCO, the World Health Organization, the World Meteorological Organization or the International Atomic Energy Agency. These organizations generally have a high proportion of people who are recruited on the basis of their scientific and technical profile (engineers, agronomists, sociologists, physicians...). But the share of those who fulfill mainly administrative tasks is difficult to assess. International organizations also rely on one-off services of experts recruited for short missions with greater scientific content. They also benefit from the contribution of "national experts" made available to them by administrations of Member States for periods ranging from 2 to 4 years. These people are not included in the available statistics. For these reasons, it is not possible to accurately assess the magnitude of scientific and technical human resources in international organizations.

However, if the statistical reality is difficult to define, there is by contrast no doubt about the issue of influence arising for a country from the presence of its nationals in international organizations, whether scientific or not. Certainly, people recruited by an international organization are at the service of its interests and missions and not those of their country of origin. National origins should not come into play, and that builds the absence of quotas by nationality in most organizations. In practice, however, when composing the teams, efforts are made to ensure balance and diversity of origins by country, and recruitments result from a game where national influences tend to offset each other. It would be ill-advised for an organization to allow too much room to certain nationalities (the "headquarters effect" being a significant bias[15]). But this search for a balance in recruitment in fact confirms the importance of the national factor. All countries value being well-represented by their nationals, especially in decision-making positions. Apart from pure prestige considerations, there is the belief that they can be vehicles of information and influence. When a decision is to be taken, the tendency of these actors towards national preference may ultimately tip the scales in favor of one course of action or another. A good positioning within the hierarchy may also encourage the recruitment and promotion of nationals in the institution and serve as a lever for the reproduction of influence.

National authorities implement strategies to promote candidates originating from their country, they disseminate information on vacancies in international organizations, they assist applicants in the preparation for recruitment exams and

---

[15]Agents holding the nationality of the country hosting the headquarters of an organization are generally over-represented in it: the French are in the majority at UNESCO, the Americans at the IMF, and the Belgians at the European Commission.

develop specific training programs, such as the "Junior Professional Officer" program of the Mission of international civil servants of the Ministry of Foreign Affairs in France. Access to executive positions is the strongest issue. Here diplomacy gets involved. When a compatriot is in the running, embassies are called upon to lobby the authorities of their country where they are located in order to win support at the time of selection. Positions in international scientific organizations are no exception to this practice of lobbying by embassies.

3.2.3.2 Scientific Expertise as Means of Influence

Expertise in an international context is considered, in general, as a vector of influence and an expression of soft power. This is particularly evident when a country, a developed one in general, is asked by another, less developed, to train its staff, support reorganization of its judicial system or its hospital activity. The country who sells its expertise promotes its technical standards, promotes its working methods and disseminates its good practice: it hopes for positive payoffs in the longer term. This is particularly true for scientific expertise, whose links with foreign policy are particularly visible in multilateral diplomacy which deals with global issues. Similar to the case of permanent staff of international organizations, the presence of nationals in international expertise panels is a lever of influence.

We focus here on expertise to serve public decision-making, which can be defined as "an expression of knowledge formulated in response to a request from those who have a decision to make, knowing that this response is intended to be integrated into a decision-making process" (Roqueplo 1997). The expert "knows but does not decide". He is a skilled-in-the-art person, recognized by his peers, who draws from his research, experience and professional practice capacities that allow him to answer the questions put to him by the one who "decides, but does not know". The contribution of the expert is an intellectual one: he undertakes the state of the art or provides advice, without concealing uncertainty or controversial elements.[16] Whether individual or collegial, scientific expertise is part of the "evidence-based policy" process, that is to say, of public decision based on the existence of empirical evidence. Originally applied in the field of medicine before being extended to other areas, this method starts from the premise that as the amount of information available increases, it is more and more difficult for practitioners to make an appropriate use of it. The quantity of scientific publications is growing, sometimes under the influence of fashion, most often due to a need to increase knowledge: "What is at stake in the debate is the opportunity for policy makers to make the best use of available knowledge, regardless of the place they intend to give this knowledge in their decision making" (Laurent et al. 2009, p. 859).

---

[16]The French Academy of Sciences promoted in 2012 a charter of expertise, which states that "the expert report mentions questions that the state of available knowledge does not enable to answer with sufficient certainty. It outlines controversies related or not to uncertainties and indicates any divergent opinions expressed in the committee of experts. It recalls that the points settled with sufficient certainty are based on the state of scientific knowledge at the time".

International negotiations on topics where the light shed by science is essential are a specific type of context where scientific expertise relating to public decision resonates with diplomatic issues. These negotiations take place between representatives of several states; they are conducted under the auspices of existing international conventions or help with preparing the adoption of new conventions. Experts are hired by governments and international organizations, and they are involved both in the preparation of negotiations and the definition of "national positions" and in the design of regulatory measures to be applied. Expertise is an indispensable tool for the exercise of multilateral diplomacy, which has a scope of intervention, especially on environmental issues, which has developed considerably over the last 30 years. At the interface of science and public decision-making, scientific expertise has influence. In the last chapter of this book we will examine in detail how scientific expertise is part of the diplomatic game in the example of international climate negotiations.

Finally, one should not omit that, next to the expertise at work within the framework of multilateral diplomacy, there is a whole area of expertise which is part of market activities and subject to intense competition between states. This international expertise is an economic issue of great importance: it "represents first a market, and a market that is growing fast" (Tenzer 2008, p. 25). This market, which is about 500 billion euros a year, brings together a highly heterogeneous demand and supply. Requests come from states or local communities, development supporting bodies or NGOs. It covers areas as diverse as defense, education or construction of infrastructure. It takes place in a bilateral or multilateral framework, for example through tenders of the United Nations. In response to this demand, the supply of international expertise comes from a wide variety of public bodies (ministries and agencies that place their experts in international organizations or respond to calls for tenders), enterprises or private offices (often turned to consulting and engineering). A sign of the recognized importance of international expertise in the diplomacy of influence is that it exists in all major countries entities within the Ministry of Foreign Affairs or working in close coordination with it, such as France Expertise International, the Department for International Development in the United Kingdom, the Crown Agents foundation in the United States, or CANADEM in Canada.

3.2.3.3 Hosting International Scientific Organizations and Major Research Infrastructures

Hosting scientific structures such as the permanent secretariat of an international agreement or the headquarters of an international organization is a source of appeal and influence. These permanent structures are forums of reflection and debate, through the conferences and various meetings which are held regularly and the comings and goings of experts and researchers involved in it: hosting cities and countries benefit from a positive image effect, like Vienna (International Atomic Energy Agency), Paris (UNESCO) or Bonn (IPBES[17]). For these reasons, the

---

[17]IPBES is the Intergovernmental Science-Policy Platform on Biodiversity and Ecosystem Services.

choice of the location of head office is subject to competition. When a country is a candidate to host a science-based organization, diplomatic structures are mobilized into lobbying tasks, which, as we have already noted, are one of the many facets of science diplomacy (see Box 3.1).

The choice of location of large research infrastructures follows the same logic of influence. But in this case the scientific challenge is direct and immediate. Such heavy equipments (dedicated to fundamental physics or astronomy, calculation tools or library resource centers...) are places where researchers from around the world concentrate. For the country which hosts them, they are a major lever for attracting grey matter and acquiring or strengthening their status on the world stage as a "country of science". Selecting their site of establishment is a matter of intense diplomatic negotiations. This illustrates the role of diplomacy in support of science, which we will detail later. Finally, the benefits in terms of consumption spending and new jobs in the local economy are additional advantages, as illustrated by the example of the European Organization for Nuclear Research (CERN). With an annual budget of about 1 billion Swiss francs, half of which consists of wages, CERN is an important support to the Geneva economic area. Orders placed with local firms amount to approximately 80 million Swiss francs a year and represent a work force of a 1000 people. Maintenance and development of equipment needed by CERN enable these companies to improve their skills in areas such as computer science or vacuum and refrigeration technologies (Bourquin 2004).[18]

Box 3.1 The choice of the headquarters of the General Secretariat of the Intergovernmental Science-Policy Platform on Biodiversity and Ecosystem Services (IPBES)

IPBES is an international group of experts on biodiversity. Mirroring the Intergovernmental Panel of Experts on Climate Change (IPCC), IPBES is an interface between scientific expertise and public decision. Under the aegis of the UN, its role is to assess biodiversity and ecosystems and contribute to the development and implementation of policies relating to them. In 2016, 126 countries were members of IPBES.

IPBES was officially launched at the second session of a plenary meeting held on 21 April 2012 in Panama. The choice of the country hosting the headquarters was made on this occasion. Five countries were candidates: France, Germany, India, Kenya and the Republic of Korea. Germany won: IPBES formally established its headquarters in Bonn.

The weeks before the vote were marked by discreet but intense lobbying, especially made by the embassies of candidate countries to the relevant authorities of their country of residence, most especially directed to delegates

(continued)

---

[18]Michel Bourquin, former Rector of the University of Geneva, chaired the CERN board from 2001 to 2003.

> **Box 3.1** (continued)
> that these countries planned to send to the Panama meeting in order to represent them. Korea promised a contribution of $3 million per year for the functioning of the Secretariat and for funding activities for the benefit of developing countries. The French bid had focused, in addition to a prestigious location (Palais de Chaillot), on the very favorable environment due to the presence in Paris of 167 diplomatic missions and 70 international and regional organizations including several involved in the field of biodiversity. An exceptional grant of $5,00,000, spread over the first 3 years of operations and a $3,00,000 support for the funding of science completed the French offer. In its candidacy Germany highlighted the expected synergies of locating in Bonn where 18 UN agencies were already present; above all, it offered an annual financial commitment of $2 million for operations and $6.5 million per year for capacity building.
>
> Ninety-two countries participated in the vote, the bid with the fewest votes being removed before each new round. Applications from India, then France and Kenya were successively rejected. In the final round, Germany won by 47 votes against 43 to the Republic of Korea.

Science diplomacy is one of the forms by which diplomacy expresses itself, and similarly takes place in the context of relations between sovereign nations. The purpose of this chapter was to demonstrate that it is a national issue. If the relationship of science to the concept of the nation can be questioned when science is considered from a cognitive point of view, it should instead be brought to the forefront when research policies are at stake. At the crossroads of the world of science and the world of diplomacy, a country's science diplomacy is underpinned by three principles of action, which are its fundamentals: attraction—cooperation—influence.

Viewed through the prism of interstate relations, global science is in the grip of a dual logic of competition and cooperation. But the definition of national science diplomacy and the mix of these two approaches vary across countries. The United States, with the attractive force of scientific capital they are known for, may have less need for international cooperation than, for example, Japan, which suffers from a deficit of openness to the world and must compensate for its lack of scientific attractiveness by building upon international cooperation networks.

At the European Union level, the cooperative thinking is at the forefront, as suggested by the role assigned to the European Research Area in the political integration of the Old Continent: while large European countries, considered individually, compete to attract talents, they also increasingly engage together in research programs funded by the EU. India and China—emerging scientific powers in Asia, whose expatriate brainpower resources are important—stride into the logic of competition and develop their appeal primarily towards their scientific diaspora.

The next chapter will be devoted to the diversity in national approaches to science diplomacy.

References

Amable, B. 2003. Les systèmesd'innovation. In *Encyclopédie de l'innovation*, ed. P. Mustar and H. Penan, 367–382. Paris: Economica.

Auriol, L. 2010. *Careers of Doctorate Holders: Employment and Mobility Patterns*. OECD Science, Technology and Industry Working Papers 2010/04. OECD Publishing. doi:10.1787/5kmh8phxvvf5-en.

Bazouni, Y. 2005. *Le métier de diplomate*. Paris: L'Harmattan.

Bourquin, M. 2004. Grâce au CERN, la région bénéficie d'une meilleure visibilité internationale. *Campus no.* 7. University of Geneva.

Choudaha, R., and C. Li. 2012. *Trends in International Student Mobility*. WES Research & Advisory Services. 21 p. http://www.wes.org/RAS

Copeland, D. 2011. *Science Diplomacy: What's It All About?* CEPI-CIPS Policy Brief (13): 1–4.

Delcorde, R. 2005. *Les mots de la diplomatie*. Paris: L'Harmattan.

Edgerton, D. 1997. Science in the United Kingdom: A Study in the Nationalization of Science. In *Science in the Twentieth Century*, ed. J. Krige and D. Pestre, 815–838. Amsterdam: Harwood Academic Publishers.

Franzoni, C., G. Scellato, and P. Stephan. 2012. *Foreign Born Scientists: Mobility Patterns for Sixteen Countries*. Working Paper No. 18067, National Bureau of Economic Research. 25 p.

Halary, C. 1994. *Les exilés du savoir—Les migrations scientifiques internationales et leurs mobiles*. Paris: L'Harmattan.

Lafforgue, L. 2005. Le français, au service des sciences. *Pour la Science*, 329. http://www.pourlascience.fr/ewb_pages/a/article-le-francais-au-service-des-sciences-20519.php

Laurent, C., et al. 2009. Pourquoi s'intéresser à la notion d' "evidence-basedpolicy"? *Revue Tiers Monde* 4 (200): 853–873.

Loriol, M., F. Piotet, and D. Defolie. 2008. *Le travail diplomatique. Un métier et un art*. Research Report to the French Ministry of Foreign Affairs, Laboratoire Georges Friedman—UMR8593. 131 p.

Ministry of Foreign Affairs—Directorate General of Global Affairs, Development and Partnerships. 2013. *Science Diplomacy for France*, 17 p. http://www.diplomatie.gouv.fr/fr/IMG/pdf/science-diplomacy-for-france-2013_cle83c9d2.pdf

National Science Board. 2012. *Science and Engineering Indicators*. Arlington, VA: National Science Foundation. 592 p. http://www.nsf.gov/statistics/seind12/pdf/seind12.pdf

———. 2014. *Science and Engineering Indicators*. Arlington, VA: National Science Foundation. 600 p. http://www.nsf.gov/statistics/seind14/content/etc/nsb1401.pdf

Pestre, D. 2004. Le nouvel univers des sciences et des techniques: une proposition générale. In *Les sciences pour la guerre: 1940–1960*, ed. A. Dahan and D. Pestre, 11–47. Paris: Editions de l'Ecole des Hautes Etudes en Sciences Sociales.

OECD. 2010. *Ministerial Report on the OECD Innovation Strategy—Innovation to Strengthen Growth and Address Global and Social Challenges*, 27 p. http://www.oecd.org/sti/45326349.pdf

———. 2011. *International Mobility*. OECD Science, Technology and Industry Scoreboard 2011. OECD Publishing. doi:10.1787/sti_scoreboard-2011-26-en.

Piotet, F., M. Loriol, and D. Delfolie. 2013. *Splendeurs et misères du travail des diplomates*. Paris: Hermann.

Roqueplo, P. 1997. *Entre savoir et décision, l'expertise scientifique*. In collection on Sciences en questions. Paris: Institut national de la recherche agronomique. 111 p.

Royal Society and American Association for the Advancement of Science. 2010. *New Frontiers in Science Diplomacy: Navigating the Changing Balance of Power?*, 32 p. http://diplomacy.aaas.org/files/New_Frontiers.pdf

Salomon, J.-J. 1970, republished 1989. *Science et politique*. Paris: Economica. 407 p.

———. 2006. *Les scientifiques entre pouvoir et savoir*. Paris: Albin Michel.

Siegfried, A. 1950. *L'âme des peuples*. Paris: Hachette.

Suttmeier, R.P. 2010. From Scientific Tourism to Global Partnership (?): Thirty Years of Sino-American Relations in Science and Technology. In *Sino-American Relations—Challenges Ahead*, ed. Y. Hao, 23–40. Farnham: Ashgate.

Tenzer, N. 2008. *L'expertise internationale au cœur de la diplomatie et de la coopération du XXIe siècle—Instruments pour une stratégie française de puissance et d'influence*. Report to the Prime Minister. 430 p. http://www.ladocumentationfrancaise.fr/rapports-publics/084000476/

UNESCO. 2010. *UNESCO Science Report 2010—The Current Status of Science Around the World*. Executive Summary. Paris: UNESCO Publishing. 520 p.

Science in Diplomatic Apparatus: The Diversity of National Approaches

4

How is science diplomacy taken into consideration and implemented in diplomatic apparatus? To answer these questions, we turn to an examination of science diplomacy of five major European countries (France, Germany, Italy, Switzerland, United Kingdom), three developed countries outside Europe (Canada, Japan, United States), two emerging countries (China, India) and Russia.

This chapter is devoted to science diplomacy as expressed in the context of relations between states, the analysis of the multilateral approach being dealt with later. We focus here on the role that embassies play in the field through their scientific and technological networks. It is important to first define them before examining their contribution to the science diplomacy of major countries.

4.1 Scientific and Technological Networks

4.1.1 Science Counselors

Diplomatic missions of major countries employ permanent staff specialized in dealing with files related to science and technology. In the traditional and almost universal vocabulary of embassies, they fall into the category of "attachés": they are "science attachés". But other names are in use, such as science advisor, science officer, science diplomat or science counselor, the latter expression fitting well enough with the "conseiller pour la science et la technologie" that is used in the vocabulary of French or Swiss diplomacy.[1] Throughout this book we use the term "science counselor" to refer generically to executive managers of the scientific and technological networks of diplomacy.

---

[1]However, attachés did not disappear from the vocabulary of French embassies, where an "attaché pour la science et la technologie" is acting as a deputy science counselor.

© Springer International Publishing AG 2017
P.-B. Ruffini, *Science and Diplomacy*, Science, Technology and Innovation Studies, DOI 10.1007/978-3-319-55104-3_4

At the embassy, the science counselor is the specialist of science, research and innovation issues. In varying degrees depending on the country, he typically fulfills four main missions:

– Collect and analyze information. A science counselor is permanently in a watch mode in order to identify and analyze scientific advances and R&D and innovation strategies of players in his country of residence. He informs his capital through diplomatic telegrams and delivers more widely science and technology watch newsletters to research centers, innovation structures and enterprises in his country of origin.
– Facilitate contacts between the communities of researchers of the two countries. The science counselor promotes the mobility of doctoral students and researchers between the country where he works and the country he represents. He supports existing scientific cooperation, provides assistance for the creation of new partnerships by organizing expert missions and seminars on topics of strategic interest, and facilitates the development of bilateral networks.
– Promote intellectual productions originating from his country and enhance its scientific and technological image. The science counselor organizes events highlighting the contributions of leading researchers and encourages their participation in the exchange of ideas. Its promotional role does not exclude lobbying authorities of the country of residence, when their support is sought for important decisions in intergovernmental organizations, as already seen in the previous chapter.
– Organize the reception of official delegations. The science counselor prepares and supports visits of ministries and executives of research institutes. More broadly, he facilitates communication between the government authorities in charge of research and innovation in both countries.

In his field of action, the science counselor informs, promotes, influences. He serves the interests of his country and promotes its values. These general objectives are shared by all countries. But particularities of both the country of origin of the counselor and the country of his assignment color his daily work. This is what we will try to highlight in our comparative review.

4.1.2 Networks

Science counselors and their teams constitute "science and technology networks" which are a subset of embassy networks. Our goal is to give an overview of the size of the main national scientific networks and of the geographical coverage they represent.

Few countries have a consistent science and technology network: the number of those who can afford significant dedicated human resources in their embassies is about ten. Few countries have opted for a strong presence, both by the number of counselors and by the number of countries where they are deployed. More often, human resources are found only in a small number of capitals. Other smaller countries have a case-by-case presence or assign research and innovation files to agents dedicated only part time to them, being in charge of other responsibilities at the embassy (academic or cultural cooperation, economic affairs. . .). In other cases, the counselor may be in charge of neighboring countries, for example, deal with Canadian affairs from a location in Washington. On the ground, the diversity of situations makes the analysis difficult. An additional difficulty is that the data provided by countries on the extent of their scientific network are often incomplete and heterogeneous. Finally, situations can change quickly. For all these reasons, the statistics gathered here should be considered with caution. The figures presented are however useful to value the orders of magnitude and make some comparisons.[2]

Data on science and technology networks are completed in this chapter with indications of what we will call "track II networks": overseas representative and administrative coordination offices of research institutions. On behalf of them, these networks perform tasks similar to the ones performed by the diplomatic network: search of contacts, support for partnerships, science and technology watch, promotion of national expertise . . . These representative offices should not be confused with research centers that national agencies can settle abroad. The figures we present below only take into account these administrative structures (representative offices, liaison offices, etc.) and do not include research units abroad, such as for example those of the German Max Planck Institute, or the *Unités mixtes des instituts français de recherche à l'étranger* (UMIFRE) in the field of social sciences.[3]

The rest of this chapter is devoted to a detailed examination of the 11 countries considered. For each of them, we will first recall what marks the originality of the international science and research strategy and then study how diplomatic tools support it. A comparative synthesis will conclude.

---

[2]This little studied topic was discussed in a report commissioned in 2008 by the German ministry of foreign affairs; it was prepared by the Wissenschaftszentrum Berlin für Sozialforschung (WZB) and issued in April 2009 under the title *Aufgabenkritische Analyse deutscher Aussenwissenschaftspolitik (AWP) in sechs ausgewählten Zielländern.* An article was adapted from it by Flink and Schreiterer (2010). A comparative analysis of national scientific networks is also available in Berg L.-P. (2010).

[3]Likewise, we do not take into account foreign bases intended to promote student mobility, such as those of Campus France or the German DAAD. We do not consider either establishments primarily devoted to the dissemination of the language and culture, such as the Goethe Institute or the Confucius Institute.

4.2 Science Diplomacy of Major European Countries: France, Germany, Italy, Switzerland and the UK

4.2.1 France

France adopted in 2009 a "National Research and Innovation Strategy" (Ministry of Higher Education and Research 2010). This general report, updated in 2013–2014, designed the long-term prospects of the insertion of the world's fifth scientific power in the European and global context. In order to develop scientific production, in which the country often excels, and stimulate innovation, where there is room for improvement, three priority areas were identified: health, care, nutrition and bio-technology; environmental urgency and eco-technology; information, communication and nanotechnology. On the international stage France aims to pursue three goals side by side: expanding its participation in the European research area in the context of the Horizon 2020 Framework Programme; intensifying its relations with high scientific and technological profile countries (Japan, South Korea…); and maintaining its historical commitment to the developing world. This general report was complemented in 2012 by a strategic agenda for 2020 (Ministry of Higher Education and Research 2013).

The National Research and Innovation Strategy of France found its continuation in the field of science diplomacy with "Science Diplomacy for France", a policy paper published in 2013 by the Ministry of Foreign Affairs. This document mainly focused on analyzing and illustrating the influential role of science in France's foreign policy and proposed guidelines for the future. It was prepared in the course of discussions on scientific diplomacy that had been underway for several years, another outcome being in 2010, with the creation of the position of an Ambassador delegate for science, technology and innovation at the Ministry of Foreign Affairs.[4]

France created its first science attaché position in 1963 in Washington (Haize 2012, p. 109). The involvement of embassies in supporting scientific cooperation has developed from 1969 (Cour des Comptes 2013, p. 74). In 2015, there were 9 counselors and 72 attachés (31 attachés for science and technology, and 41 attachés for university and scientific cooperation) covering 54 countries. The geographical distribution is given in Table 4.1 below. Much of human resources are concentrated in Europe, with a presence in 26 countries (including six outside the European Union).

The network is placed under the authority of the Ministry of Foreign Affairs and steered by the sub-directorate of Research and Scientific Exchange (Cultural, Academic and Research Cooperation Directorate). In the nine most important countries for scientific cooperation, there is a service at the Embassy of France dedicated to science and technology, which reports directly to the Ambassador and

---

[4]A position held since its creation by physicist Catherine Bréchignac, Permanent Secretary of the French Academy of Sciences. Her mission is to promote French scientific and technological excellence and to support the country's research strategy internationally.

Table 4.1 French counselors and attachés for science and technology in the world (2015)

Region[a]	Counselors	Attachés	Total
Africa/Middle East (7)		7	7
North America (2)	2	10	12
Latin America (6)		7	7
Asia (11)	3	18	21
Europe[b] (26)	5[c]	28	33
Oceania (2)		2	2
Total	10	72	82

Source: Ministry of Foreign Affairs
[a]Number of countries in parenthesis
[b]Including Russia
[c]Including a counselor (Brussels) funded by the Ministry of Higher Education and Research

is led by a counselor with a team of several science attachés, young international volunteers and administrative staff. In these major countries,[5] the workforce dedicated to science and technology issues ranges from 3 (Spain) to 25 people (US). The other areas of cooperation (cultural, educational, linguistic, academic. . .) fall under the counselor for cultural cooperation. In other countries, attachés for science and technology are part of a cooperation and cultural action service alongside with other attachés dealing with other main areas of cooperation. In many countries, science attachés are also involved in academic cooperation (they are "attachés for university and scientific cooperation").

The French track II network is particularly significant. Table 4.2 summarizes the distribution of foreign sites of five large public organizations that are invested partly or entirely of a research mission. The overseas offices of these organizations are not research centers but support offices interfacing with the key players of the host country mainly in areas of French excellence (nuclear, space, health sectors. . .). With two-thirds of their settlements located in the Africa/Middle East and Asia regions, the *Institut Pasteur* and the *Institut de Recherche pour le Développement* reflect the importance attached to a ground presence in France's traditional regions of influence.[6]

4.2.2   Germany

Research and innovation strategy of Germany and its international scope are described in two policy documents from the Federal Ministry of Education and Research (BMBF): the "High-Tech Strategy", released in August 2006 and updated in July 2010 under the title "Ideas. Innovation. Prosperity. High-Tech Strategy for

---

[5]Canada, China, Germany, India, Japan, Russia, Spain, United Kingdom, United States of America.
[6]With ten regional offices all located in developing countries, the *French Agricultural Research Centre for International Development* (CIRAD) is another vector of French influence in the South.

Table 4.2 Representative offices abroad of various research-related French organizations (2015)

Region	CEA	CNES	CNRS	Institut Pasteur	IRD[a]	Total
Africa/Middle East	2		1	10	10	23
North America	1	1	1	1	1	5
Latin America			1	2	6	9
Asia	4	2	4	7	4	21
Europe[b]	6	2	1	6	1	16
Oceania				1	1	2
Total	13	5	8	27	23	76

CEA Alternative Energies and Atomic Energy Commission, *CNES* National Center for Space Studies, *CNRS* National Center for Scientific Research, *IRD* Research Institute for Development
Source: Related organizations
[a]Excluding overseas departments and territories of France
[b]Including Russia

2020" (Federal Ministry of Education and Research 2010); and the "Strategy for the Internationalization of Science and Research", published in 2008 and updated in 2014 (Federal Ministry of Education and Research 2014), which sets out for the first time the country's objectives in the international field. The ambition it straightforwardly displayed: consolidate Germany's position as the leading center for research in Europe and develop the ability to lead research and innovation strategy of the Old Continent as a whole.[7]

To apply these guidelines at the international level, emphasis is given to increasing the visibility of the country and its attractiveness for foreign researchers. The major research organizations have implemented programs to attract PhD students and young researchers. In 2013, 30,000 researchers and academics around the world received support from German financing bodies. This policy is supported by a major budgetary effort: in 2013, R&D spending in Germany amounted to more than 80 billion euros, or 2.85% of GDP. Strategic choices of Germany also result in identifying partner countries with the exercise of influence and cooperation are now considered as priorities: Brazil, China, South Korea, India, the latter having already Germany as second partner for scientific co-publications, behind the United States.

The strategy for the internationalization of science and research of 2008 brought Germany into an assumed science diplomacy approach: an evaluation of foreign experiences was commissioned,[8] it was decided to set up abroad "German Houses of Research and Innovation" and to strengthen the scientific network by recruiting local staff. In 2013, the German scientific network consisted of 25 counselors at embassies in 20 countries, which forms a relatively compact network. It shows in particular no presence in Oceania. With three counselors in Brussels, Germany

---

[7]"Germany wants to become the driver of European strategy development in research and innovation policy", Federal Ministry of Education and Research (2008), p. 27. The official communication introduces also the "High-Tech Strategy" as a "model for Europe". See for example the brochure *Germany Inspires Innovation—Welcome to Europe's leader in Science* of the Federal Ministry of Education and Research.
[8]See note 2 above.

Table 4.3 German science counselors in the world (2013)

Region[a]	Counselors
Africa/Middle East (2)	2
North America (2)	5
Latin America (3)	3
Asia (7)	7
Europe[b] (6)	8
Total	25

Source: Embassy of Germany in Italy
[a]The number of countries is between brackets
[b]Including Russia

indicates the importance attached to the European Research Area. And with one counselor in Warsaw and two in Moscow, it confirms its historical position as a bridge between East and West of Europe and its influence on the eastern part of the region.

Table 4.3 shows the breakdown by major region of German science counselors.[9] These are sent to embassies for given durations through three channels: some are directly sent by the ministry of research, some others belong to the ministry of foreign affairs personnel, and others are agents of the ministry of research seconded to the ministry of foreign affairs and affected by the latter at embassies. It should be noted that science counselors at embassies are never academics or researchers, which seems to be explained by a proud tradition of independence from the state apparatus.

Adding to this conventional device, a strong track II network is characteristic of the German presence abroad (Table 4.4). The Deutsche Forschungsgemeinschaft (DFG), which finances German research, is present in five foreign countries.[10] The three other main organizations that are involved in public research have also representations abroad:

- The Fraunhofer-Gesellschaft, a very large application-oriented research organization, has seven representative offices and four liaison offices abroad[11];
- The Helmholtz-Gemeinschaft, another German community of research institutes, is represented in Beijing, Brussels and Moscow;
- The Leibniz-Gemeinschafthas its own representative office in Brussels.

---

[9]The table does not include locally recruited employees.

[10]With offices in Moscow, New Delhi, New York, Tokyo and Washington, and a Sino-German Center for Research Promoting in Beijing.

[11]In China, India, Indonesia, Japan, Russia, South Korea and United Arab Emirates. Fraunhofer also has liaison offices in Austria, Brazil, India and Malaysia. It also created six institutes in the United States and three in Europe (Roy 2010, 2012). S. Roy, "Le positionnement international de la Fraunhofer" note of the Service for Science and Technology of the Embassy of France in Germany, March 10, 2010; Roy S., "Partenariat à l'international des différents organismes de recherche et agences de financement allemands" note the Service for Science and Technology of the Embassy of France in Germany, January 9, 2012.

Table 4.4 Permanent representations abroad of various German organizations (2013)

Region	DFG	Fraunhofer Gesells.	Helmholtz Gemeins.	Leibniz Gemeins.	Total
Africa/Middle East		2			2
North America	2				2
Latin America					
Asia	2	6	1		9
Europe[a]	1	2	2	1	6
Oceania					
Total	5	10	3	1	19

Source: Related organizations
[a]Including Russia

Finally, the German Houses of Research and Innovation created from 2009 are a unique tool of the science diplomacy of Germany. These locations, which are intended to bring under one roof the representations of research organizations, innovation agencies and universities, are in Moscow, New Delhi, New York and SãoPaulo.[12]

4.2.3 Italy

Italy shows a paradoxical image. In the light of European standards, scientific research is notoriously underfunded: R&D spending represents 1.26% of GDP, against 2.01% for the EU average (2013). Adding to severely constrained budgets of public research, there is an overall weakness of private research, which suffers from the industrial structure of the country, relatively poor in large enterprises. This lack of investment results in a damaging consequence: brain drain. According to OECD figures, about 300,000 higher education graduates live abroad. It is estimated that around 6000 researchers leave the country each year.[13]

The roots of this exodus lie in the inability of the university and research system to guarantee the best researchers attractive working conditions and career

---

[12]For example, the German House of Research and Innovation in São Paulo includes the representations of the following institutions: Alexander von Humboldt Foundation (AvH); CLIB 2021—Cluster Industrial Biotechnology eV; Academic Exchange Service (DAAD); German Research Foundation (DFG); Freie Universität Berlin; Georg-August-UniversitätGöttingen, Technical University of Munich (TUM); UAS 7—German Universities of Applied Sciences University Alliance Metropolis Ruhr (UMAR); TASK—The Centre of Competence for Soil, Groundwater and Site Revitalisation of the Helmholtz Centre for Environmental Research (UFZ); WestfalianWilhelms-University of Münster.

[13]25% of them go to the USA, 20% to Great Britain, 16% to France (where they generally represent the biggest part of foreign researchers at the national Center for Scientific Research, with about 14% of the total). See *La Repubblica*, March 5, 2013.

advancement. Yet, according to *Science Watch* data, Italy succeeds in reaching eighth position worldwide for the impact of its publications, which amount to about 50,000 each year: space science, clinical medicine and physics are among the areas of Italian excellence. These results are based largely on the work of expatriate researchers: according to the *Top Italian Scientists* ranking, published by the *Virtual Italian Academy* (a network of academics and Italian researchers in the UK), half of the top 100 Italian researchers (as measured by the *h-index* of publications) live and work outside Italy.

Italian science diplomacy should be approached through this particular lens. The research and innovation strategy of the country was presented in the 2011–2013 National Research Programme, authored by the Ministry of Instruction, University and Research (MIUR). But there is no single official document dedicated to the internationalization of research and innovation, and no explicit strategy for science diplomacy. Yet, evidencing that such a strategy is not born only the day it is named as such, the country has been preparing for science diplomacy for a long time. In 1999, the Italian Ministry of Foreign Affairs issued a remarkably documented book on the role of the science attaché, a book that is hard to find the equivalent in other countries (Ministero degli Affari Esteri 1999). In 2010, the Department transformed its organization with the creation of a Directorate General for the promotion of the "countrysystem". A unit for bilateral and multilateral science and technology cooperation was created, thus giving better visibility to the scientific cooperation network abroad.

The Italian scientific network consists of 25 *addetti scientifici* in 20 countries (Table 4.5). With the exception of the United States and China, which each have three science counselors, representation is usually provided by one person at the embassy. Adding to this, some Italian Cultural Institutes abroad may host "science experts" (e.g., in San Francisco), with duties which do not differ fundamentally from those of science counselors. In countries where there is no staff dedicated to scientific and technological cooperation, related files are usually assigned to the economic and commercial counselor.

Table 4.5 Italian science counselors in the world (2015)

Region[a]	Counselors
Africa/Middle East (3)	3
North America (2)	5
Latin America (3)	3
Asia (5)	7
Europe[b] (4)	4
Oceania (1)	1
Internat. organizations: OECD, UN	2
Total	25

Source: *Innovitalia* (Ministry of Foreign Affairs, MIUR)
[a]Number of countries in parenthesis
[b]Including Russia

Italian science counselors are academics and researchers, and more rarely officials from the Ministry of Research. No career diplomat is to be found among them. The Ministry of Foreign Affairs ensures the recruitment and payment of these officers, while maintaining a formal and informal consultation with the ministry in charge of research for all the strategic issues related to the network: geographic coverage, recruitment profiles, etc. It also coordinates the network. Efforts to create a truly interactive work between science counselors and their parent ministry were undertaken from 2010.

In spite of limited resources and after a significant contraction in 1993, the size of the Italian network remained stable and even increased slightly over the past 15 years. Following an agreement signed in December 2011 between the Ministry of Foreign Affairs and the Italian Space Agency, two positions of space attachés were created within diplomatic and consular missions in Washington and Brussels.

Italian science counselors perform conventional tasks of monitoring and support for cooperation. But, given the Italian peculiarity of brain drain, they are also mobilized to maintain links with the community of science expatriates and, if necessary, to give rise to the creation of an association of Italian researchers in the country where they operate. Italy shows much interest in its scientific diaspora and tries to use it as an asset by connecting it to the national production and research system. For this purpose, the *Innovitalia platform*, a "social network of Italian brains", was launched in 2012 jointly by the Ministry of Foreign Affairs and the Ministry of Research. The scientific network of the peninsula is in charge of promoting this means of promoting research among expatriate researchers which, it seems, has no equivalent in other countries.

4.2.4 Switzerland

Switzerland performs remarkably in the field of research and innovation. Since 2008, its R&D expenditure has been exceeding the symbolic threshold of 3% of GDP, the same one the European Union has set its goal. Relative to its population, national scientific production stands at 3.2 publications per thousand inhabitants, which places the country at the top of the world ranking together with Finland. For innovation, the country is among the top runners: first place in 2015 at the *Innovation Union Scoreboard* and at the *Global Innovation Index* calculated by INSEAD and the World Intellectual Property Organization. In 2015, it ranked third in the *World Competitiveness Scoreboard* of the International Institute for Management Development and ranked fourth according to the *Global Competitiveness Index* of the World Economic Forum.

The high quality of education and research institutions in Switzerland explain these dominant positions at the top of the world hierarchy. The strategic watchword is simple: "To maintain this level of performance to meet the growing competition and socio-economic changes" (Federal Council 2012).[14] The Federal Council chartered the strategy to follow: to affirm the status of Switzerland as an internationally recognized scientific and competitive economy; to ensure its leading position in the promising new areas of research and in the space industry; to strengthen cooperation between science and the economy ... These are all goals for a leading country which cares to remain a leader by consolidating its positions. The message of the Federal Council is remarkable in that it explicitly addresses the role of science and technology networks in the implementation of this strategy. Together with framework agreements concluded with high scientific profile countries and joint programs with priority countries, this network is one of the three tools of bilateral cooperation.[15]

The Swiss scientific network is under the dual supervision of the Federal Department of Foreign Affairs and the State Secretariat for Education, Research and Innovation. People who run it fall under either of these departments: they are therefore either academics or career diplomats. In 2015, there were 20 people stationed in 20 countries.

This network is relatively old, as the first counselor was installed in Washington in 1958. After position openings in Tokyo (1986) and Brussels (to the European Union, 1990), it expanded greatly from the mid-1990s. It is since that time that science diplomacy has been truly considered. In 1993 further reflection was developed on the future of the network, and an interdepartmental agreement was set up. Expansion decisions were made. By the turn of the century, embassies of Paris, Pretoria, London, Singapore, Beijing, Moscow and Rome were provided with a counselor in charge of science (Wisard 2010). In 1999, the Federal Council discussed for the first time the role of science counselors in a four-year message on the Promotion of Education, Research and Technology for the period 2000–2003. It announced that additional human and financial resources would be allocated to the development of the network (*ibid.*) (Table 4.6).

The network was enhanced in the 2000s with the creation of five "knowledge hubs" of great originality. Under the name Swissnex (in which *nex* means both *nexus*, *next* and *next to*), these infrastructures are platforms supporting the international cooperation of higher education and research-involved companies. They develop contacts in their host countries, organize events, facilitate partnerships, and more

---

[14]The Federal Council delivers messages every 4 years on the international promotion of education, research and innovation. In 2010, the Federal Council issued its vision of the international strategy in the fields of education, research and innovation. See *Internationale Strategie der Schweiz im Bereich Bildung, Forschung und Innovation*, www.sbf.admin.ch/bfi-international.pdf.

[15]It is worth mentioning a remarkable feature of the Swiss scientific diplomacy: scientific neutrality has been "put to work to the benefit of the political neutrality of Switzerland during the Cold War". This strategy guided "the participation of Switzerland in the establishment of three international organizations for scientific research at European level: the European Organization for Nuclear Research (CERN) in 1953, the European Space Research Organization (ESRO) in 1961 and the European Molecular Biology Conference (EMBC) in 1969" (Strasser and Joye 2005, p. 61).

Table 4.6 Swiss science counselors in the world (2015)

Region[a]	Counselors
Africa/Middle East (1)	1
North America (2)	2
Latin America (3)	3
Asia (5)	5
Europe[b] (8)	8
Oceania (1)	1
Total	20

Source: State Secretariat for Education, Research and Innovation
[a]Number of countries in parenthesis
[b]Including Russia

generally in key countries promote the excellence of education, research and innovation in Switzerland. They are fed mainly by private financing (Dorsaz and Marmier 2010). As said by Christian Simm, director of Swissnex San Francisco, "a Swissnex platform is not a consulate, not an economic development office and not a startup, a little of each of all these all in one" (Guillaume 2011). Twenty people in all work in the settlements of Boston (2000), San Francisco (2003), Shanghai (2008), Bangalore (2011) and Rio (2014). In 10 years, the Swissnex network has forged a strong identity. It has become a reference in the international arena, and a model for science diplomacy of other countries: it inspired Germany's opening in 2009 of its first house for research and innovation.

4.2.5 United Kingdom

With 14% of the most cited research papers in the world, the United Kingdom ranks second behind the United States. Relative to the number of researchers, published articles and citations are higher than in the US, Japan or Germany (Department of Business, Innovation and Skills 2011a). Four British universities are among the top ten in the "Shanghai ranking". Since its inception, the Nobel Prize has been awarded more than one hundred times to British researchers. The United Kingdom has been and remains a great country of science. In international competition, it has an additional asset: that of being at the forefront for the inclusion of science in diplomacy.

In the UK, the portfolios of higher education and research are held by the Department of Business, Innovation and Skills (BIS). This department was created in 2009 and its extensive expertise also includes business and enterprise, trade and innovation. The overall vision of the research and innovation policy is outlined in the text *Innovation and Research Strategy for Growth* (Department of Business, Innovation and Skills 2011b). Innovation is at the heart of the strategy. In this guidance text, the words "science diplomacy" are not being used. But science diplomacy is fully present in the words and actions of the Foreign and Commonwealth Office (FCO), the ministry of foreign affairs of the United Kingdom.

Interdepartmental coordination and strategic guidance of the country's international commitment in science and innovation is provided by a steering committee, the *Global Science and Innovation Forum*. "Chief scientists" are associated to this steering body. The UK is one of the few countries that have empowered scientific leading figures to advise policymakers at the highest levels of the state. Since 1964 a Government Chief Scientific Adviser has been serving the Prime Minister and the members of the Cabinet.[16] He is appointed by the government and his key role is to bring science into the public decision process. The Chief Scientific Adviser is the head of a small team, the *Government Office for Science*, whose role in steering the international strategy of the country is central. This unit published in 2012 an assessment report on how the Foreign and Commonwealth Office uses science and engineering, which is a first source for understanding the British science diplomacy. While confirming that science diplomacy is a fully embraced approach in the UK, this uncompromising report recognizes without fear or favor that there is scope for improvement (Government Office for Science 2012).

There are also Chief Scientific Advisers at most government departments: in 2015, there were 23. This has been the case at the Foreign and Commonwealth Office since 2009. The position is staffed by a renowned scientist, whose part-time role includes providing advice to government officials and to interface between the capital city and science counselors at embassies.

Within a large diplomatic network (more than 14,000 people in around 270 sites), the overseas network of UK science diplomacy has been organized since 2000 in a *Science and Innovation Network* (SIN). This network works for all government departments and public agencies. It was established in 2000 by the Foreign and Commonwealth Office (FCO). It has been co-managed and co-funded since 2008 by the Department for Business, Innovation and Skills and by the FCO. It spreads over 28 countries and employs 90 agents, with a reasonable balance between expatriates and locally recruited staff. The former are academics or researchers, or diplomats who may have a science background by education although this is not a rule. Local staff are recruited on the basis of their knowledge of the country, and they generally have a scientific background. These science officers work together with UK Trade and Investment, a government agency which promotes British companies in foreign markets and facilitates the establishment of foreign companies in the UK.

Table 4.7 below shows the geographical breakdown of all personnel employed by the SIN network in 2015.[17] It is worth noting that almost half of human resources are concentrated in Asia, and slightly more than a quarter in Europe. Latin America holds a marginal position, with a presence only in Brazil.

The British science and technology network is human resource-efficient. It draws on hubs for regional coordination: for example, the Paris office coordinates the Western Europe region, and the Berlin office is responsible for Central and Eastern Europe. While excluding any increase of the resources devoted to the

---

[16]Sir Mark Walport, a physician, has served in this position since 2013.

[17]Irrespective of qualifications: counselors and attachés, but also assistants and translators. Some jobs may be only part-time. The figures in this table cannot be compared to those given about other countries in this chapter, which relate only to the expatriate workforce.

Table 4.7 Total
workforce of the *Science
and Innovation Network*
(2015)

Region[a]	Number of staff
Africa/Middle East (6)	9
North America (2)	18
Latin America (1)	3
Asia (7)	37
Europe[b] (12)	21
Oceania (2)	2
Total	90

Source: British Embassy in France
[a]Number of countries in parenthesis
[b]Including Russia

network, the 2012 Government Office's evaluation report recommended maximizing the number of countries covered by relying on expatriate science officers for regional coordination, and relying on local personnel for presence in the field: this pattern applies in particular to Southeast Asia or North America.

Overseas offices of UK Research Councils in Beijing, Brussels, New Delhi and Washington DC are part of the British track II network. They bring additional support to bilateral cooperation with countries or regions considered strategic.[18]

4.3   Science Diplomacy of Developed Non-European Countries

4.3.1   Canada

Eighth economy in the world, Canada is a country that has strong foundations in research. Canada is proud to be the first among G7 countries for the number of papers published per capita. The country has a strategy for science and technology, which is described in the official document, *Mobilizing Science and Technology to Canada's Advantage* (Canada's New Government 2007). Bearing the signatures of the Ministry of Industry—of which depends the research sector—and the Ministry of Finance, the strategy assigned to science and technology an explicit mission: to strengthen the competitive advantage of the country on the world stage. This fell within the comprehensive economic plan *Advantage Canada—Building a Strong Economy for Canadians*, launched in 2006, and which aimed to make the country a world leader through science and technology. At the international level, research policy is characterized by an openly commercial orientation. This is confirmed by the departmental organization, which showed from 1993 to 2015 a *Department of Foreign Trade and International Affairs*.[19] Unique among the countries under

---

[18]There are also the resources of the British Council, which has a budget dedicated to science of £8–10 million and a team of four scientific advisers (Embassy of France in the United Kingdom 2011).

[19]The ministry of foreign affairs was renamed *Global Affairs Canada* in 2015.

Table 4.8 Canada's innovation network abroad (2014)	Region[a]	Agents
	Middle East (Israel)	1
	North America (United States)	8
	Latin America (3)	3
	Asia (6)	7
	Europe[b] (13)	14
	Oceania (1)	1
	Total	34

Source: Canada Trade Commissioner Service
[a]Number of countries in parenthesis
[b]Including Russia

review, this department was co-led by a minister of foreign affairs and a minister of international trade, the latter being responsible for science and technology international cooperation.

To implement this strategy, Canada has a science and technology network—officially named an "innovation network"—which in 2014 spread over 25 countries (Table 4.8). In key locations abroad, it employed 34 agents with a science-technology-innovation profile. There were also trade commissioners in more than 150 cities abroad.

There are two types of counselors in Canadian Embassies: "S&T counselors", with an academic science/research profile, and "Trade Commissioners" to which science and technology files may be entrusted. These two figures are present in the most important countries. Trade commissioners are most numerous. On the ground, whether they are science counselors or trade commissioners in charge of science affairs or the promotion of a particular industrial sector, all are expected to develop a broad and pragmatic approach, combining science, innovation and industrial issues: all of them work for the promotion of an integrated model of international trade.

The Canadian scientific network is thus clearly mobilized to support Canada's economy and business. We should point out this strong originality of the Canadian model: while the policy internally supports fundamental as well as applied research, as it is in every country, innovation promotion and competitiveness are the hallmarks at the international level, where corporate values and success in the marketplace prevail.

4.3.2 Japan

Recent developments in the global science landscape are a challenge for Japan. Against a background of demographic decline, the archipelago is facing the rise of its Asian neighbors in its natural area influence. Globally, its contribution to the production of knowledge is eroding in relative share.[20] In this context, the insufficient opening of the country to the rest of the world is problematic (Sunami et al. 2013). At a time when global research and science are more and more a matter of networks, great research infrastructures, multinational programs and brain circulation, improving the participation of Japanese researchers in this global game has become a cardinal duty. Public authorities are aware of the need to open the national research community more to the outside world.

These data direct the definition of science diplomacy, through which the country hopes to find appropriate responses. The opening in this direction occurred in 2008, with the publication of a founding text entitled *Towards the Reinforcement of Science and Technology Diplomacy* (Council for Scientific and Technology Policy 2008) issued by the Council for Scientific and Technology Policy (an interdepartmental body that sets the country's science policy). The debate which was opened at the highest levels of the state upheld the view that science and technology diplomacy ought to be one of the vectors of the revitalization of Japan, particularly in order to strengthen the R&D system by an inflow of foreign resources (The Cabinet 2010).[21] In late 2010, the Council for Science and Technology Policy issued the *Japan's Science and Technology Basic Policy Report*, which prefigured the *4th Science and Technology Basic Plan*, launched the same year by the Japanese government (Council for Science and Technology Policy 2010). This master plan was the first to explicitly address science and technology diplomacy, to which it assigned two major objectives: to promote Japan's own assets in R&D in the global competition, and to strengthen its influence in emerging countries and developing countries, with a particular emphasis towards East Asian countries.[22] The 4th Plan assigned Japan to work for regional cooperation by taking the initiative of the creation of an "East Asia Science and Innovation Area".[23]

Two departments are responsible for translating these guidelines into action: the Ministry of Foreign Affairs (MOFA) and the Ministry of Education, Culture, Sports, Science and Technology (MEXT). The latter implements its international research programs through two funding agencies: the Japan Society for the Promotion of Science (JSPS) and the Japan Science and Technology Agency (JST). A particularly significant initiative was born in 2008 thanks to the coordination between the two ministries: the *Science and Technology Cooperation on Global Issues* program. An aspect of this program consists in sending Japanese researchers

---

[20]For 22 research disciplines, the share of China and South Korea in world publications has increased since the early 2000s, while that of Japan has declined (Sunami et al. 2013).

[21]*Ibid.* See also *The New Growth Strategy: Blueprint for Revitalizing Japan*, issued by the Cabinet in June 2010. http://www.meti.go.jp/english/policy/economy/growth/report20100618.pdf.

[22]In that respect, a "Strategic Program for Building an Asian Science and Technology Community", operated by the Japan Society for the Promotion of Science (JSPS) was launched in 2006, with the support of the *Special Coordination Funds for Promoting Science and Technology*, and covering the period 2006–2010.

[23]The *e-Asia Joint Research Program*, which is at the heart of this initiative, was launched in Singapore in 2012.

Table 4.9 Japanese science counselors in the world (2012)

Regionᵃ	Counselors
Africa/Middle East (1)	1
North America (2)	6
Latin America (1)	1
Asia (7)	7
Europeᵇ (6)	6
International organizations (6)	6
Total	27

Source: Embassy of Japan in Paris
ᵃNumber of countries or organizations in parenthesis
ᵇIncluding Russia

Table 4.10 Permanent representations of various organizations abroad (2013)

Regionᵃ	JSPS	JST	NEDO	Total
Africa/Middle East (2)	2			2
North America (USA)	2	1	2	5
Asia (4)	2	2	2	6
Europeᵇ (4)	4	1	1	6
Total (11)	10	4	5	19

Source: Organizations' official websites
ᵃNumber of countries in parenthesis
ᵇIncluding Russia

to developing countries in order to conduct their joint research activities with local researchers (*Dispatch of S&T Researchers Program*). Another program (*Science and Technology Research Partnership for Sustainable Development*—SATREPS) promotes joint research with developing countries on global issues.

The Japanese scientific network includes 27 counselors (see Table 4.9). It is notably present in Asia, with recently created positions of science counselors in Vietnam and in Jakarta (Permanent Delegation to ASEAN). The vast majority of scientific counselors are selected from MEXT staff. Academic as well as career diplomat profiles are exceptional.

Japan stands out by the importance of its track II network, which significantly raises the representation of its scientific interests abroad. Taken together MEXT funding agencies (JSPS and JST) have 14 offices in 10 countries (Table 4.10). The New Energy and Industrial Technology Development Organization (NEDO), which implements international programs of the Ministry of Economy, Trade and Industry in applied research areas is represented in five foreign countries.

4.3.3   United States

In the global hierarchy, the United States is and remains the number one country in the world of science. With their 228 Nobel prizes in all scientific disciplines, the United States leaves the United Kingdom (77), Germany (58) and France (34) far behind. More than one-third of those Nobel prizes have been awarded to American scientists born outside the United States or who received their PhD outside the United States. But far from undermining the American performance, this confirms the force of attraction of this country's knowledge production models.[24]

The United States has a global science policy. With Europe and China challenging its supremacy, the first country in the world for science and knowledge production aims to maintain its rank. For this, it is critical that what makes the country's strength continues to work: the ability to attract and integrate the grey matter from around the world. Maintaining a continuous and regular inflow of human capital devoted to research is one of the major issues of the American science diplomacy. A second and more directly political objective was the use of knowledge and science in the new definition of the foreign policy set by the Obama administration. These recent trends were recorded in two major documents issued under the titles "International Science and Engineering Partnerships: A Priority for US Foreign Policy and our Nation's Innovation Enterprise" (National Science Board 2008) and "Global Science Program for Security, Competitiveness, and Diplomacy Act" (House of Representatives 2012).

The current period is marked by a striking renewal of science diplomacy. But the premises date back to the end of World War II, when it became clear in the United States that the role that science would be called to play in foreign policy would no longer be limited to its links with the military. Since the 1960s, the US science diplomacy had more or less regularly grown in power. In the 1990s, the relative decline in the interest borne by foreign policy to science and technology issues was only transitory. A report of the National Research Council identified issues of science, technology and health in almost all the objectives of the country's foreign policy (National Research Council 1999). This influential study paved the way for the appointment of a science and technology adviser to the Secretary of State the following year.

But September 11, 2001 brought a halt to momentum in science diplomacy. Its aftermath was marked by a strong hardening of conditions of entry into the territory for certain nationalities of researchers. Then the approach swung with the return to power of the Democrat administration: the doctrine of smart power gained currency. In the exercise of this smart power, in which diplomacy should seek to define the best combination of instruments in each case, those falling under the power of influence through knowledge assets and advances in research occupy a large place. The *Global Science Program for Security, Competitiveness, and Diplomacy Act* of 2012 was the legislative expression of the new presidential guidance. Two strategic goals were displayed: national security and support to competitiveness. To that end, the draft law provided for a higher share of science in foreign policy. It was considered important that scientists, engineers and innovators—an underutilized asset, according to the text of the bill—be put towards promoting the diplomatic objectives of the country.

---

[24]Data from 2012.

Since the Roosevelt administration, at the White House there has been a Science Advisor to the President. In 1976, the Congress established the Office for Science and Technology Policy, headed by the presidential science advisor.[25] The Department of State also benefits from scientific advice. Since 2000 there has been a Science and Technology Adviser to the US Secretary of State. His task is to raise the scientific input in the Department of State. He leads a team responsible for "bringing to bear the vast resources of the U.S. science community to build science capacity across the Department".[26] An important aspect is to ensure the link with the network of embassy counselors. The position has been occupied since its origin by recognized personalities from the world of research.[27]

In the US organization charts, the steering of science and technology international policy is operated by the White House Office for S&T Policy and by the State Department, in which the Bureau of Oceans and International Environmental and Scientific Affairs is a significant source of proposals, and supervises the scientific network abroad.

With the creation of science advisor positions, substantial effort has been made in the US to make policy makers aware of science issues. Two specific programs complete these provisions:

- The *Science & Technology Policy Fellowships* program

This program was established by the American Association for the Advancement of Science. It enables scientists and engineers, recent doctoral graduates or experienced professionals (about 300 annually) to be recruited to work at the interface of science and public action in federal agencies, offices of the Congress or in scientific professional societies. Career opportunities are open to them in the fields of diplomacy, security and development. The program should allow for incorporating scientific and technical expertise in all phases of preparation, implementation and evaluation of foreign policy issues (issues of non-proliferation and health security, environmental issues, international trade and multilateral cooperation). Today, each year, it enables the allocation of about 30 researchers to scientific expert positions to the State Department, for a period of 1–2 years. After that, many of them join the State Department and serve in Washington or in the scientific network at embassies (Fedoroff 2009).

- The *Jefferson Science Fellowship program*

---

[25]John Holdren, a specialist in environmental sciences, was nominated to this position in 2014.

[26]The Office of the Science and Technology Adviser to the Secretary (STAS), http://www.state. gov/e/stas/.

[27]In September 2015, the post was confided to Vaughan Turekian, former director of international relations of the American Association for the Advancement of Science.

Created in 2004, this program is intended for experienced academics, researchers and engineers, and enables them to work for the State Department or the United States Agency for International Development (USAID) as scientific advisors on foreign policy issues. They are assigned for 1 year and are paid by their university, according to the principle of the sabbatical leave. Then they return to their home institution but continue to work for the State Department as expert consultants for a period of 5 years. The revival of science diplomacy in 2010 resulted in increased financial support for this program, and in an increase in the presence of scientists in the Department of State.

The first scientific office of the American diplomatic network was established in 1947 at the Embassy in London. In 1950 the National Security Council published its first directive on scientific information gathering abroad.[28] The same year the Berkner report was published (Department of State 1950). This was the first comprehensive study about how to consider science in the State Department. The report advocated for setting up a science office in the Department, and also to develop a network of science counselors in fifteen US embassies in non-Communist countries. Following these recommendations, the number of counselors jumped to ten. But in 1956, this network was eliminated. Adding to budgetary reasons was a sense of malaise due to the inadvertently made public desire of the CIA to entrust science counselors with an intelligence mission (Lexow 1966). The first Soviet space successes in 1957 strung the Americans badly, and they restored their network. In 1958, seven counselors were appointed. In 1960 their number climbed to 14, then to 23 in 1965, in 17 embassies (*ibid.*).[29]

Today, officers of the American scientific network are called *Environment, Science, Technology and Health (ESTH) officers*. The backbone of the network consists of 12 regional hubs, allowing some cross-border issues (such as environmental issues) to be apprehended on a regional scale. These regional officers interact with national officers in charge of the bilateral relationship.[30] In 21 other countries, ESTH officers deal only with cases concerning their country of residence. These personnel are career diplomats. In total, the presence of science counselors and science officers spans 33 countries and four international organizations. In 146 other countries, covering virtually the entire American diplomatic network, science files are supported on an ad hoc basis by agents dealing mainly with economic and political issues (Table 4.11). For what concerns the track II network,

---

[28]Valuable information on the implementation of scientific advisors network in US embassies are given by Wilton Lexow, head of the CIA's Applied Sciences Division, in an internal memo written in 1966 and declassified in 1994 under the title *The science attaché program*. See also Commission on the Organization of the Executive Branch 1955.

[29]At the same time, 25 countries used a science counselor in their embassy in Washington (1960).

[30]These regional hubs are located in: Accra, Addis Ababa, Amman; Gaborone; Lima, San Jose; Astana, Bangkok, Kathmandu; Budapest, Copenhagen; Suva. ESTH officers network's activities are regularly described in reports that the Department of State publishes on its website: http://www.whitehouse.gov/administration/eop/ostp/sciencediplomacy.

Table 4.11 The American scientific network (2013)

Region[a]	Science counselors	Science officers	Environmental hubs	Total
Africa/Middle East (7)		3	4	7
North America (Canada)	1	–	–	1
Latin America (4)	2	4	2	8
Asia (11)	2	12	3	17
Europe[b] (9)	2	13	2	17
Oceania (1)	–	–	1	1
International organizations (4)	3	3	–	6
Total	10	35	12	57

Source: Own compilation
[a]Number of countries in parenthesis
[b]Including Russia

we note finally that the *National Science Foundation* has a permanent representation in Paris, Beijing and Tokyo.

The American presence in the field is completed par two programs unique to that country:

• The *Embassy Science Fellows* program

Under this program (2002), scientists can be employed at US embassies. They act as consultants and advisors working with ESTH officers. In 2010 the program was strengthened, and the initial 3 months assignment duration was extended to 3 years.

• The *Science Envoys* program

This program consists in sending eminent American scientific figures on mission to foreign countries. These special envoys travel as private persons and report to the White House, the State Department and the American scientific community. From 2009 to 2015, 13 special envoys have been commissioned to identify areas of mutual interest and potential partnership.

4.4 Science Diplomacy of China, India and Russia

4.4.1 China

China has engaged in a rapid and ongoing development of its research capacity, which has the objective of moving the country from the stage of workshop of the world to the stage of laboratory of the world. The National Medium and Long Term Program for Science and Technology (2006–2020) and the prospective analyses of the Chinese Academy of Sciences (Lu 2010) have outlined this new and gigantic leap forward. R&D spending, which has grown roughly 20% per year since the early 2000s, should represent 2.5% of GDP in 2020.

Communist China began in the 1970s to normalize its diplomatic relations through the channel of scientific exchanges: the "Shanghai communiqué" signed in 1972 by the President of the United States Richard Nixon and the President of the Chinese Communist Party Mao Zedong provided for such exchanges. Seven years later, with the establishment of diplomatic relations between the two countries, a formal agreement on scientific and technological cooperation was signed. This remarkable episode, which features prominently in the world history of science diplomacy, opened the way for the normalization of relations between Chinese scientists and the international scientific community. The 9th workshop on science diplomacy, held in 2011 in Beijing, was the occasion to point to the progress achieved: in 2010, China had established cooperation relations for science and technology with 152 countries and regions worldwide and signed 104 scientific cooperation agreements, representing an active partnership with 97 countries or regions (Embassy of France in China 2011).

The entire scientific and technological policy is placed under the control of the Ministry of Science and Technology, the powerful MOST. The Chinese Academy of Sciences, the China Association for Science and Technology, the Chinese Academy of Engineering and the National Natural Science Foundation of China also contribute to it. MOST works out plans, supervises and controls budgets. It is also the largest research funding agency in the country, through numerous programs under its responsibility. This department ensures the strategic management of international scientific cooperation, through its own programs or programs of other departments (*ibid.*).[31]

The Ministry for Science and Technology controls over the Chinese scientific network. MOST (Department of Personnel, Department of International Cooperation) recruits, supervises and pays counselors for science and technology at embassies (Embassy of France in China 2011).[32] In 2012, the network employed 141 people in 69 diplomatic missions and 46 countries (Embassy of France in China 2011). The geographical breakdown is given in Table 4.12 below.

Network agents are mostly MOST officials, notably from its international department. Some may come from other ministries (culture, information) or from science and technology departments of provinces. More rarely, they come from research institutes or universities. No career diplomats are found among them. They have more often a scientific background and are also chosen for their foreign language skills.

---

[31]Between 2006 and 2010, the Chinese international science and technology cooperation program has funded 1728 projects, totaling 4.38 billion yuan (486 million euros). Over 3600 foreign researchers have received aid from the Chinese government, via four specifically dedicated major national programs.

[32]http://www.most.gov.cn/eng/organization/Mission/index.htm.

Table 4.12 Science and technology network of China (2012)

Region	Number of countries
North America	7
Africa/Asia/Oceania	13
Europe[a]	26
Total	46

Source: Embassy of China in Italy
[a]Including Russia

Analyzing geographical priorities is instructive. Chinese science diplomacy obviously seeks to build close ties with major areas of science in North America, Europe or Japan. But the country's comparative advantage lies more in its relations with developing countries, where its strategy of influence is very marked. This results in numerous cooperation and training programs: 310 scientific seminars were hosted between 2001 and 2011 by China, which supported financially the invitation of more than 5700 scientists and representatives of developing countries (Embassy of France in China 2011). The Forum on China-Africa Cooperation, established in 2000, is a privileged channel through which China expresses its influence on this continent. It is also worth noting that the World Academy of Sciences for the advancement of science in developing countries (TWAS) in 2012 elected a Chinese president in the person of Bai Chunli, president of the Chinese Academy of Sciences.

4.4.2 India

Like China, India has been experiencing fast growth since the early 2000s. But unlike its powerful neighbor, India is not an economic giant. With over a billion people, its GDP does not exceed that of France, and no remarkable model of industrial development is at work there. But along with China, India is among the countries where R&D spending is made at a steady pace. The existence of an important scientific diaspora is another common point between the two countries: India is facing brain drain, and nothing indicates that the outflow could be quickly stopped or reversed.

India has great ambitions in the field of science and innovation. It ranks ninth for the number of publications, among which physical sciences and life sciences stand out. When addressing the 100th annual session of the Indian Science Congress, which was held from 3 to 7 January 2013 in Calcutta on the theme "Science for Shaping Future of India", the Prime Minister set forward the goal of doubling the country's global share of publications by 2020. This would raise it to 7% and allow the country to enter into the top five of science countries in the world (Foot note) Source: Embassy of France in India. India adopted the necessary means for achieving its ambition with the announcement of its "Science Plan, Technology & Innovation Policy 2013", a plan which provides for the doubling of the research and innovation budget in 5 years (Government of India—Ministry of Science and Technology 2013).

International cooperation is essential to achieving this objective and has a high margin for development. In the wake of the pioneering agreement signed between India and the Soviet Union in 1972, the country has increased bilateral agreements, concluded to date with 73 countries. However, India has prioritized scientific collaborations with Asian countries since the early 2000s and still suffers from lack of openness to Western Europe.

The Ministry of Science and Technology is in charge of international cooperation, which includes the preparation and implementation of bilateral and multilateral agreements, as well as the responsibility for scientific activity at international organizations. This duty is conducted in close collaboration with the Ministry of Foreign Affairs. But in contrast the number of Indian embassies involved in it is low. There are four science counselors in the Indian network (*Science Wings in Indian Missions* in Berlin, Moscow, Tokyo and Washington). Adding to this, there are technical liaison officers in Austria, France, the United Kingdom and the United States with missions focusing on space, nuclear or defense issues.[33]

4.4.3 Russia

Russia today barely maintains the glorious scientific legacy of the Soviet Union from which it derived. The time is long past when the deeds of space exploration pioneers were enough to tell the world of the talent of Soviet researchers and engineers and the success of a military and industrial organization heavily irrigated with science. With the economic and social stagnation of the 1990s and the brain drain that ensued, the country's research system has paid a high price for the disappearance of the Soviet Union. Today, despite a steady improvement in overall economic conditions since the early 2000s, Russia has not recovered its rank of major scientific power. For 10 years, domestic spending on R&D peaked at 1.2% of GDP, and the relative share of the country in world publications has declined.

Can science be a source of soft power for Russia? Basic sciences—mathematics, physics, some specialties of chemistry—remained areas of excellence. But the national research system is struggling to reform itself. These last years, it moved closer to international standards (creation of a status of "national research university", recourse to international expertise for the evaluation of programs, initiatives to promote innovation. . .). But state and bureaucracy remain pervasive: the share of state-funded R&D is about 80%, an unusually high percentage compared to countries with which Russia could be compared. The 2013 reforms increased political control over the powerful Russian Academy of Sciences. Paternalism and the weight of hierarchy are slow to fade within research laboratories. All this does not make Russia an attractive location for researchers from other countries (Dezhina 2012). In this context, the primary objective of science diplomacy is to restore the international image of Russian research.

---

[33]Official website: http://www.dst.gov.in.

The Russian scientific network abroad has a trompe l'oeil aspect. Its most visible part consists of "Russian centers for science and culture". These institutions are coordinated by *Rossotrudnichestvo*, an agency under the Ministry of Foreign Affairs. Its main missions are to maintain Russian influence in the Commonwealth of Independent States and, more broadly, to promote cultural and educational exchanges with the many countries where it is represented.[34] In 2012, the Agency had 59 centers for science and culture over the world. Although the word "science" is in their name, these centers have only limited involvement in science and technology, mainly through the organization of exhibitions and conferences. They are mostly dedicated to the dissemination of Russian language and culture: they are cultural centers, in the usual diplomatic sense. Their role is also to strengthen links with the Russian expatriate community.

The scientific network within Russian diplomatic structures is more limited. A decree from 2003 created the "attachés for science and technology". Since 2011, these science counselors have been appointed and administered by the Ministry of Higher Education. Their number reached 15 in 2015. These agents rarely have an academic profile: engineers or personnel from the Ministry of Foreign Affairs or the Ministry of Research are found among them. At big Russian embassies, their duties are not always comparable to that described in this chapter for other countries under review. Often, science and technology files are processed by the economic adviser.

This overview focusing on the most important scientific countries and on some emerging countries does not do justice to other countries of a smaller size or with a more recent economic development. Some of them are worth mentioning here for the attention they bring to the representation of science and research in their diplomatic network: Austria, Finland, Hungary, Israel, South Korea, Turkey … The latter country officially announced in 2012 the creation of science counselor positions in its consulates of San Francisco and Boston and in its embassies in Germany and Japan. Not to mention the European Union, which installed scientific counselors within its delegations in Brazil, the United States, Ethiopia (where the African Union is based), Japan and Russia. The European Union does not use in its communication the term "science diplomacy", but benefits from influential tools which the European Research Area and the Framework Programs for Research and Technological Development that may include researchers from non-member countries.

4.5   The Diversity of National Models

While pursuing the same general goals (attraction, cooperation, influence), the countries under review are facing specific problems and the scientific diplomacy they implement is marked by their history and their institutions. As a conclusion,

---

[34]In official terms, *Rossotrudnichestvo* is "the Federal Agency for the Commonwealth of Independent States, compatriots living abroad, and international humanitarian cooperation".

we would like to propose a comparative review by placing these countries' experiences in relation to each other. This tentative synthesis will be conducted from three questions: How is the strategy of science diplomacy defined and managed? What choices in terms of human resources are made to implement it in diplomatic networks? What are the key characteristics allowing for building typical national models?

4.5.1   Science Diplomacy Is Unevenly Assumed

All the countries we examined have a strategy for research and innovation. But in official documents describing it, the importance given to the international dimension of the strategy varies significantly. Similarly, not all countries invest in science diplomacy to the same extent. A first category of countries consists of those who use the term "science diplomacy" in their political and institutional vocabulary: this is the case of France, Germany, Japan, the USA and the United Kingdom. While recognizing the vocabulary in the communication of the ministries in charge of it, these countries have appropriated the concept of science diplomacy to varying degrees, and make the necessary efforts of reflection and organization for implementing what they see as a strategic issue.

Anglo-Saxon countries are the pioneers in this field. The leading position of the United States lies not only in their prior presence on the ground, but also in the conceptual renewal provided by the concepts of soft power and smart power, which were designed in the US. The Cairo speech was a milestone signing the revival of science diplomacy as desired by the Obama administration. But reflection on science diplomacy and dissemination of the vocabulary in the United States also owes much to the initiatives of the American Association for the Advancement of Science, which created in 2008 a Center for Science Diplomacy and launched in 2012 *Science & Diplomacy*, the first journal devoted entirely to this topic. The UK also played a pioneering role. This country stands at the forefront, with a scientific network that was set up at the end of the Second World War.[35] The United States and the United Kingdom are the countries where science diplomacy is best understood intellectually. Science ranks among the priorities of their diplomatic strategy. They are also differentiated by opening their networks to global issues, environmental ones in particular: American counselors are *Environment, Science, Technology and Health* officers, and it is expected from the staff of the British network SIN that they exercise effective lobbying on these major issues (Department for Business, Innovation and Skills—Foreign Commonwealth Office 2012).

---

[35]In 1941 the first British science officer was appointed abroad: the grandson of Charles Darwin became director of the Central Scientific Office in Washington. In 1942, Joseph Needham, a member of the Royal Society, was appointed head of the British Scientific Mission in China (Royal Society and AAAS 2010).

To a lesser degree, other countries have embraced the discourse and the goals of science diplomacy. Since 2008, Japan has been officially using the vocabulary and has set strategic guidelines. This is also the case of Germany. France also takes place in this group, as witnessed by the publication in 2013 of the official report *Science Diplomacy for France*. Other countries however have not, or not yet, used this vocabulary—which does not mean that they ignore science diplomacy. The example of Switzerland is most instructive in this regard. Science diplomacy is absent in words, but firmly rooted in actions. In its "Message on the Promotion of Education, Research and Innovation", not only did the Federal Council set the objectives of the country regarding the international policy of research and innovation, but it also set the role that the diplomatic tool should play in the field and indicated which changes should be made in the network in order to achieve the policy objectives. For other countries, the link between foreign policy and the world of science and research may be less explicit. Italy, Russia and China fall into this category.

4.5.2   Diplomatic Networks and Their Geographical Coverage

Diplomatic networks are field tools of science diplomacy. Data on them are incomplete and heterogeneous. Our view is that the least questionable comparisons can be made by looking at the number of foreign countries where the presence of at least one full-time science counselor is provided.[36] Three countries are in the lead according to this indicator of geographic coverage: France (54 countries), China (46) and the US (33). The United Kingdom (29) and Canada and Japan (25) follow. Germany, Switzerland and Italy are present in about 20 countries.

For all countries under review, being present in countries of high scientific level (G7) or rapid growth (emerging countries) is essential. Switzerland, Canada or Japan have very similar location choices: their science counselors are present in major European countries, the USA and China. This is also the case of the UK, which downsized its network in the 1990s and decided to focus on the most important countries. But priorities to have locations in the ten or so major countries in science and technology do not exclude particular tropism: we saw for example that Japan attaches great importance to Asia, with a presence in seven countries.

Some countries cover with their network much larger geographical areas. Although traditionally showing a strong partiality for the developing world, France has a global approach: historically, this country stands out by the extent of its diplomatic and linguistic influence (with its embassies, cultural centers, French schools, *Alliance Française* offices...). Note however that the French scientific network in 2013 entered into a downsizing phase, in the context of an overall reduction of public spending.

---

[36]We hesitate to include Russia in the leading group, given the very special nature of its scientific network.

Among the countries we examined, France, Japan and Germany are distinctive by the size of their track II network. The case of Japan is particularly striking: its track II network accounts for half of the total number of facilities which the country possesses abroad.

4.5.3 Steering and Coordination

How are the guidelines of science diplomacy set? How are scientific offices steered in diplomatic networks? How does coordination take place between international strategies of research institutions from the same country?

In all countries we examined, two departments are on the front line: the Ministry of Foreign Affairs and the Ministry in charge of research. As an almost general rule, the scientific and technological network is set up, financed and supervised by the Ministry of Foreign Affairs. It is this department that decides on the size and geographic distribution of facilities, and on hiring and assigning staff members—whether or not diplomats. This is not surprising insofar as these personnel are based in embassies and their work consists of bilateral cooperation and diplomacy of influence. But in all countries, the Foreign Ministry is working more or less closely with the "thematic" ministry, the one in charge of research policy.

Two countries have the highest osmosis at the ministerial level between foreign affairs and research policy: the United Kingdom and Switzerland. In the UK, the SIN network is under the ultimate supervision of the Foreign and Commonwealth Office but is co-funded and co-managed by the Department of Business, Innovation and Skills. It is not uncommon that these two departments jointly publish guidance material or evaluation reports. In Switzerland, the State Secretariat for Education and Research and the Federal Department of Foreign Affairs jointly provide the network monitoring. The thematic department plays a particularly important role in decisions that affect the scientific network in Japan (MEXT) and even more in China (MOST).

But to be effective, coordination must not be limited to strategist ministries. In the UK, the coordination between ministries and academies is done within the "Global Science and Innovation Forum". In Switzerland, an interdepartmental "Coordinating Committee of Science and Research" (IDA-WI) was reactivated in 1993. Acting as a strategic body, this committee brings together representatives of the Federal Department of Home Affairs, the Federal Department of Foreign Affairs and the Federal Department of Economic Affairs, Education and Research (Wisard 2010, p. 26). In Japan, an inter-ministerial "Council for promoting cooperation based on S&T diplomacy" was established (Council for Science and Technology Policy 2010, p. 23): renamed *Council for Science, Technology and Innovation* in 2014, it includes four departments, but also research agencies such as JSPS, JST or NEDO. In the United States, an act of 26 March 2009 established a committee to coordinate international scientific and technological cooperation. But these examples are of limited scope: they express the need for coordination, but coordination remains more functional than strategic.

The definition of a national strategy for science diplomacy is subject, more fundamentally, to the special nature of the relationship between the world of research and that of politics. In all countries, what characterizes institutions which conduct or represent scientific activity (research institutes, universities, academies...) is their will for autonomy vis-à-vis the executive, whatever it may be. The relationships between science and politics (and therefore diplomacy) are thus subject to a fundamental tension between freedom of initiative demanded by researchers and their institutions, and the guidance set by those in charge of defining and funding the Nation's research policy on behalf of the general interest. Representation abroad of major research institutions provides a practical illustration. These institutions define their strategy autonomously. They decide to open representative offices abroad or enter into international consortiums on behalf of a vision of international development of their own, and not by reference to a unified "national science diplomacy" approach that does not exist in any country. These players' international strategies probably do not come into conflict with the overall guidelines adopted by national strategies for research and innovation—which for the most developed countries generally lead to the same recommendations (to work preferably with high science potential and rapid growth countries, to focus on health, environment, nanoscience and nanotechnology fields...). National frameworks are sufficiently open and non-binding so that each player can bring in its own international strategy.[37] And as already noted at the beginning of this book, these international practices or research stakeholders are a part of science diplomacy, although they may not express any visible interaction between science and foreign policy. They belong in a diffuse but real way to the broad area of the diplomacy of influence.

But this might result in a lack of consistency. For example, "the French research bodies (...) have their own, i.e. not joint, representative offices abroad presenting a fragmented picture of our research instruments and centers, despite our diplomatic posts' coordination efforts" (Ministry of Foreign Affairs 2013, p. 14).[38] In the UK, the Research Council UK (RCUK) created its own network, because academic researchers' priorities did not fit with that of the scientific network SIN (Flink and Schreiterer 2010).[39] Pragmatic answers are sought to the true question of coordination in the field between "networks" and "Track II networks". In most cases, non-formalized but effective working relations exist on the ground between the various representatives of national research. In the Japanese case, a liaison

---

[37]Thus, in Germany, a country where researchers are generally very reluctant to coordinate with the government, little more can be done officially than to call for all those acting internationally to "work together" (Federal Ministry of Education and Research 2008).

[38]In France, the issue of coordination of diplomatic science networks, and more generally of networks of the state and of public operators abroad, was discussed in two recent evaluation reports: the report of the General Inspectorate of Finance and the General Inspectorate of Foreign Affairs (2013), already mentioned, and the joint report of the Ministère de l'enseignement supérieur et de la recherche and the Ministère des affaires étrangères (2014).

[39]In the UK, funds for research are allocated through seven councils, federated within the RCUK.

committee of research organizations has been created in some host countries where these organizations are present (China, USA, France, Great Britain, Malaysia) (Embassy of France in Japan 2012). But the most notable initiative was taken by Germany, which gathered its research institutes in "Houses of Research and Innovation", adding to the benefits of pooling resources the benefit of high visibility thus given to national research in the host country. But this interesting experience should not be overestimated: no building as such is dedicated to these "Houses". It is rather a matter of shared office spaces, quite distinct of German diplomatic posts.

In all countries, it is ultimately out of reach and even irrelevant to achieve a kind of "integrated command" of the national science diplomacy, which would revolve around shared objectives of international initiatives of all research stakeholders. But this rapid review shows that there is room for improving the coordination of players' strategies worldwide: everywhere, "the challenge lies in an effective, recurrent and sustainable combination of bottom-up interest aggregation with strategic decision-making" (Flink and Schreiterer 2010, p. 676) and in finding an optimal fit between spontaneous bottom-up initiatives from research institutions and top-down guidelines and incentives from directly concerned ministries.

4.5.4   Human Resources for Science Diplomacy

"Smart power requires smart people" (Department of State-USAID 2010, p. XVI). Diplomacy is first a matter for diplomats. A large majority of them did not come into contact with science and research activity during their academic studies. Given the increasing technical level of many topics, diplomats consult experts and are assisted by specialized advisers. Their attitude towards issues of science—which they often perceive as complex, if not impenetrable—depends a lot on their personal sensitivity or curiosity. Still, the success of the science diplomacy approach depends on the ability of diplomats to understand the essentials of issues and diagnoses identified by science. Multilateral negotiations on global science-intensive issues such as climate, environment or human health provide an illustration.

Raising awareness of major scientific issues among diplomats is a human resources challenge. But to date, it remains largely absent from training programs offered by foreign ministries to their senior officials. Still, a majority of countries we examined seek to scientifically arm delegations participating in international negotiations. There are also interesting international initiatives. For instance, a partnership between the IPCC and the European Climate Foundation helped to hold workshops in Africa, Asia and Latin America for preparing officials of these regions in view of the annual conference (COP-15 in 2009) of the Climate Convention (Royal Society and AAAS, 2010). Also worth mentioning is the "Science and Diplomacy" program launched jointly by the AAAS and the international academy TWAS in the frame of the World Science Forum in Budapest in 2011, which is intended to raise the scientific preparation of diplomats from developing countries for their more effective involvement in major international conferences.

Another route is often taken to raise the science input in departmental organization charts and in public decision-making, especially in foreign policy: to create positions of science advisers at the government level and positions of ambassadors for science and technology in the Ministry of Foreign Affairs. The UK is the country where the practice is most spectacular. Scientific advice to government is organized on a large scale. Not only the Prime Minister but also 20 departments or public agencies have a chief scientist.[40] One of them is attached to the FCO, the British ministry of foreign affairs. In the United States there is a presidential science adviser, as well as an S&T adviser to the US Secretary of State. France and Japan have an ambassador for science and technology.[41] In the United States, England and France, these functions are performed by personalities with an indisputable scientific background. In Japan, however, the post-holder is traditionally a career diplomat.[42]

Foreign ministries of countries under review in this chapter assign the bulk of their staff directly involved in science diplomacy in the diplomatic networks. Scientific networks are run by officers who are recruited in three ways. There are career diplomats, assigned for a limited period (3 or 4 years). There is also staff seconded from another ministry (the ministry in charge of research, generally) and scholars from research institutes or universities: the duration of their assignment is comparable to the diplomats', and they are also expatriate personnel. Finally, there are agents recruited on a local contract for periods that can be much longer, and who fulfill various supervisory or administrative functions. This local staff is the "memory" of scientific networks, providing a salutary counterweight to the rapid turnover of expatriates. In addition to considering the criteria of occupational origin and recruitment modes, the share of science and technology responsibilities in different positions is worth consideration: some network officers are active in several areas, with only a part-time involvement in science-related files.

Whatever the administrative arrangements for their recruitment, officers of science and technology networks in short fall into two categories: "diplomats" and "professionals", the latter being expatriates or staff recruited locally. It appears from a field study conducted on scientific networks of 20 countries, based on data of 2008, that among counselors working full time on science-related files, 16% were career diplomats, 44% were professionals from other departments or organizations

---

[40]The number of ministry and other government body officials with scientific or engineering training is about 3500, including more than 700 areas of expertise. They gather in a *Group of Government Scientists and Engineers*.

[41]Since 1989 Australia has a chief scientist who advises the Prime Minister and other ministers. Israel has a chief scientist at the Ministry of Industry, Trade and Labor. The position of ambassador for science also exists in South Korea.

[42]There is in Japan a reluctance to entrust high-profile scientists with political or administrative responsibility. More generally, "most Japanese political leaders do not perceive S&T as a useful instrument for foreign policy. Even if they do, they rarely mention it in international fora" (Sunami et al. 2013).

and 40% were locally recruited staff. These percentages were respectively 26, 40 and 34% for counselors engaged part-time in science-related issues (Berg 2010).

The investigations that we carried out on 11 countries support these findings. Some countries do not resort to career diplomats, or only in an exceptional and marginal way: France, Italy, China and Japan are in this situation. But there are significant differences: France chooses academics, researchers or engineers,[43] while China and Japan fuel their network with administrative staff from their ministry of research. Conversely, other countries rely on a significant proportion of career diplomats (from 25 to 40%) and do not rely on professionals from other departments or organizations: the United Kingdom, Switzerland and Canada are in this situation. Two countries use a combination of diplomats and professionals in their recruitment, but in opposite proportions: Germany relies strongly on professionals from the ministry of research (but with no scholars or researchers) and only marginally on diplomats (around 10%), while the United States relies overwhelmingly on diplomats (60–70% of the total workforce), with professionals and local staff representing the remainder.[44]

Here we have the diplomat and the scientific professional, neither above the other depending on what is expected to be prioritized by the people in charge of the scientific network. In absolute terms, it may seem preferable that counselors are familiar with the world of research: having university qualification and a recognized academic background is likely to favorably predispose those who are interlocutors of the counselor in the scientific community of the host country. This profile is relevant if what is first expected from counselors is to support cooperation initiatives and to behave like ambassadors of the national scientific community in their country of residence: it is important that they have a good knowledge of the research and innovation system of the country they represent. This is the choice made by Italy or France. Some other countries, conversely, prefer to entrust these positions to diplomats without special academic or scientific experience. US ESTH officers are required to support the front-line presence of the United States on global issues. In the United Kingdom, environmental and especially climate issues are also at the forefront of concerns. It is therefore consistent to assign these political functions to agents originating from the world of diplomacy. It is all the better if in addition they can claim a scientific training.[45]

---

[43]"90% of science counselors and attachés for university and scientific cooperation come from universities and research institutes placed under the supervision of the Ministry of Higher Education and Research" (Ministère de l'enseignement supérieur et de la recherche et Ministère des affaires étrangères 2014, p. 27).

[44]This has not always been so: in the years 1950–1960, the American advisers were all senior scientists (Lexow 1966).

[45]In 2011, this was the case for about half of the personal of the British Science and Innovation network (Department for Business, Innovation and Skills—Foreign Commonwealth Office 2012).

4.5.5 Country Models

Science counselors cover a wide field, ranging from basic research to innovation and from the academic to the corporate world. But depending on countries, the center of gravity of their activity is not located in the same place. To conclude this comparative approach of diplomatic networks dedicated to science and technology, we would like to contrast national approaches according to the priorities they hold, but also more fundamentally, to the overall philosophy which underlie their approach. Several country models can be identified.[46]

4.5.5.1 Environment and Global Challenges Model
Two countries stand out through the missions they assign to agents of their network for the monitoring of global challenges, including environment. US science counselors are *Environment, Science, Technology, and Health* officers and this name clearly indicates their priority missions. Originating from diplomacy, US science counselors are mainly used for political tasks. For its part, the United Kingdom has embraced the "millennium goals" by the time of the Blair administration. British science diplomacy has been completely revised in the light of the major issues of the moment: climate, energy and other global issues are its watchwords. This British diplomacy's hallmark permeates all the activity of the network: on these issues, bilateral diplomacy supports and relays multilateral diplomacy.[47]

4.5.5.2 Trade and Business Profile
The network of Canada offers the most striking example of trade and business profile. As already mentioned, the country had a Department of *"Foreign Trade and International Affairs"* from 1993 to 2015. The Canadian model's originality is to entrust to its scientific network priority objectives of promoting innovation and enhancing the competitiveness of the national economy.

4.5.5.3 Higher Education-Research-Innovation Integrated Model
With its Swissnex platforms, Switzerland offers the most successful example of a network which at the same time supports international cooperation of colleges, universities and Swiss companies active in research. These science and technology outposts are mainly financed by the private sector, and in the most dynamic regions in the world (USA, China, India, Singapore) project an image of a country capable of effectively articulating scientific excellence and economic success.

---

[46]We leave aside China, India and Russia in these comments.

[47]The British Science and Innovation Network also has a strong orientation towards innovation. The focus on innovation appears in the name of the ministry responsible for research and supervision on the scientific network (Department of Business, Innovation and Skills) and in the title of the strategic document ("Innovation and Research Strategy for Growth"). The Science and Innovation Network also has close ties in the field with the *UK Trade and Investment* department.

4.5.5.4 Academic Model

The academic model is characterized by the use of scientific networks in prioritizing the interests of public research and cooperation projects involving research institutes and universities in their traditional mission of producing new knowledge. France, Germany and Italy are most representative of this approach. When claimed (as in the case for the France and Germany), support for technology transfer and innovation generally occupies a less important place in the duties of scientific counselors.

Two special features of the French model of science diplomacy must be raised. The first one is its disconnection from economic and trade issues. The French model is characterized by being heavily tilted towards the academic dimension, leaving behind innovation. This is indeed the international projection of what likely amounts to a national disadvantage already identified in the strategic reference text: "There is a notable discrepancy between the academic achievements of French research and their tangible benefits for society in terms of innovation and economic development" (Ministry of Higher Education and Research 2010). This national characteristic is not unrelated to the lack of private investment in R&D.

But there is also another unique feature: the unusual closeness of science and culture in French embassies. The organization of the diplomatic network reflects the major importance given to cultural cooperation. Certainly in nine major countries, science counselors and attachés work within a separate service at the embassy, referred to as the "Service for science and technology". But in these countries, international academic cooperation lies within the authority of the cultural counselor and is separated from scientific cooperation although it is functionally close to it. In 40 or so countries where there is no such thing as a science and technology service at the embassy, science attachés are placed under the authority of a "counselor for cooperation and cultural action". The creation in 2010 of the *Institut Français*, the public operator in charge of promoting French culture around the world, confirms and reaffirms the priority given to culture, in all its forms, in the strategy of influence. One could probably argue that science is part of culture at large, and that science diplomacy is one of the aspects, among others, of the diplomacy of influence. But such an organization leaves only a reduced visibility to research and innovation, even though they have been seen since the "Lisbon strategy" of the European Union as the critical variables of the advent of the knowledge society in Europe.

The positioning of science into the orbit of culture is not found in any other country, and is a further expression of the French exception. In networks of other countries, we find counselors working on both scientific and cultural files, as in some Swiss embassies for instance. But when there is no scientific service as such, foreign examples show that the most frequent option is to incorporate the science counselor into the economic mission (that is the case, for example, at US embassies) or to entrust science-related files to the counselor in charge of economic affairs (as at some Italian embassies). French diplomacy in 2012 took a marked orientation towards economic diplomacy: in this new context, it may be worth observing the experiences of countries who seek to raise synergies between research and business in their foreign policy, thus reflecting the benefits of closely articulating on the ground science diplomacy and economic diplomacy.

Dedicated to the analysis of the place of science in the diplomatic services of several countries, this chapter has allowed us to highlight the diversity of national situations. We have shown that the approach to science diplomacy differs among foreign ministries. Pioneers are the US and the UK. Germany, France and Japan are well positioned. Without formally using the vocabulary of science diplomacy, Switzerland and to a lesser extent Italy are part of the process. China, and even more Russia and India, are less advanced. Another lesson of this chapter is that the core missions delegated to science and technology networks vary across countries. The priority may be the enhancement of national public research (academic model: Germany, France, Italy) or the competitiveness of companies (trade and business model: Canada). Priorities may involve the major issues of the time (environment and global challenges model: the United States, the United Kingdom) or the international promotion of a national model articulating scientific excellence and economic success (integrated higher education-research-innovation training, such as in Switzerland).

Finally, recalling that within embassies the duty of science counselors and their teams is to facilitate bilateral contacts between research communities and to enhance the scientific and technological image of the country they represent, it is to "diplomacy for science" that this chapter was dedicated. The action of diplomats can support cooperation between researchers of different countries. But, in turn, international scientific relations can also facilitate the exercise of diplomacy or serve as its vanguard: the next chapter will be devoted to this "science for diplomacy" dimension.

References

Berg, L.-P. 2010. Science Diplomacy Networks. *Politorbis* 49 (2): 69–74.

Canada's New Government. 2007. *Mobilizing Science and Technology to Canada's Advantage*, 103 p. https://www.ic.gc.ca/eic/site/icgc.nsf/eng/00871.html

Commission on the Organization of the Executive Branch. 1955, June. *The New Hoover Commission Report on Intelligence Activities*. Report to Congress, Intelligence Activities, Superintendent of Documents, US Government Printing Office.

Council for Scientific and Technology Policy. 2008. *Towards the Reinforcement of Science and Technology Diplomacy*. http://www8.cao.go.jp/cstp/english/doc/s_and_t_diplomacy/20080519_tow_the_reinforcement_of.pdf

Council for Science and Technology Policy. 2010. *Japan's Science and Technology Basic Policy Report*. http://www8.cao.go.jp/cstp/english/basic/4th-BasicPolicy.pdf

Cour des Comptes. 2013. *Le réseau culturel de la France à l'étranger*, 160 p. http://www.ccomptes.fr/Publications/Publications/Le-reseau-culturel-de-la-France-a-l-etranger

Department of Business, Innovation and Skills—BIS. 2011a. *International Comparative Performance of the UK Research Base*, 80 p.

———. 2011b. *Innovation and Research Strategy for Growth*, 162 p.

Department for Business, Innovation and Skills—Foreign Commonwealth Office. 2012. *Science and Innovation Network: Annual Report 2011–2012*.

Department of State. 1950. *Science and Foreign Relations*. Department of State Publications 3860, May.

Department of State-USAID. 2010. *Leading Through Civilian Power—The First Quadrennial Diplomacy and Development Review*. Washington, DC: Executive Summary.

Dezhina, I. 2012. *The Russian Science as a Factor of Soft Power*. Russian International Affairs Council, June 21. http://russiancouncil.ru/en/inner/?id_4=515#top-content

Dorsaz, P., and P. Marmier. 2010. La nouvelle diplomatie scientifique de la Suisse et le modèle Swissnex: l'exemple de Boston après 10 ans. *Politorbis* 49 (2): 57–62.

Embassy of France in China. 2011. Focus en chiffres sur la diplomatie scientifique chinoise. *ADIT—Bulletin Electronique Chine*, 107, September 19.

Embassy of France in Japan. 2012. *Diplomatie scientifique au Japon: état des lieux*, memo of the Service for Science and Technology, June 12.

Embassy of France in the United Kingdom. 2011. Le British Council et la diplomatie scientifique. *ADIT—Bulletin électronique Royaume-Uni*, 112, November 17.

Federal Council. 2012, February 22. *Promotion of Education, Research and Innovation for 2013–2016*. http://www.sbfi.admin.ch/org/01645/index.html?lang=en

Federal Ministry of Education and Research. 2008. *Strengthening Germany's Role in the Global Knowledge Society—Strategy of the Federal Government for the Internationalization of Science and Research*, 29 p. http://www.bmbf.de/pubRD/Internationalisierungsstrategie-English.pdf

———. 2010. *Ideas. Innovation. Prosperity—High-Tech Strategy for 2020*, 19 p. http://www.bmbf.de/pub/hts_2020_en.pdf

———. 2014. *International Cooperation—Action Plan of the Federal Ministry of Education and Research (BMBF)*. Bonn: Federal Ministry of Education and Research.

Fedoroff, N. 2009. Science Diplomacy in the 21st Century. *Cell* 136: 9–11.

Flink, T., and U. Schreiterer. 2010. Science Diplomacy at the Intersection of S&T Policies and Foreign Affairs: Toward a Typology of National Approaches. *Science and Public Policy* 37 (9): 665–677.

Government of India—Ministry of Science and Technology. 2013. *Science, Technology & Innovation Policy*, 22 p. http://www.dst.gov.in/sti-policy-eng.pdf

Government Office for Science. 2012, September. *Science & Engineering Assurance Review of the Foreign and Commonwealth Office*, 27 p.

Guillaume, M. 2011. Diplomatie scientifique—Comment la Suisse est devenue championne. *L'Hebdo*, September 6. http://www.hebdo.ch/comment_la_suisse_est_devenue_championne_119351_.html

Haize, D. 2012. *L'action culturelle et de coopération de la France à l'étranger: un réseau, des hommes*. Paris: L'Harmattan. 285 p.

House of Representatives. 2012. *Global Science Program for Security, Competitiveness, and Diplomacy Act* (H.R. 6303). http://www.gpo.gov/fdsys/pkg/BILLS-112hr6303ih/pdf/BILLS-112hr6303ih.pdf

Inspection générale des finances and Inspection générale des affaires étrangères. 2013. *Mission d'évaluation de l'organisation et du pilotage des réseaux à l'étranger*, 40 p.

Lexow, W. 1966. *The Science Attaché Program*. http://www.cia.gov/library/center-for-the-study-of-intelligence/kent-csi/vol10no2/html/v10i2a02p_0001.htm

Lu, Y. 2010. *Science and Technology in China: A Roadmap to 2050—Strategic General Report of the Chinese Academy of Science*. Beijing and Heidelberg, NY: Science Press and Springer. 138 p.

Ministère de l'enseignement supérieur et de la recherche and the Ministère des affaires étrangères. 2014. *La coordination de l'action internationale en matière d'enseignement supérieur et de recherche*, 142 p. http://www.ladocumentationfrancaise.fr/rapports-publics/

Ministero degli Affari Esteri. 1999. *Scienza e tecnologia italiane all'estero—Il ruolo dell'Adetto scientifico*. Roma: Marchesi Grafiche Editoriali S.p.A, 275 p.

Ministry of Foreign Affairs—Directorate General of Global Affairs, Development and Partnerships. 2013. *Science Diplomacy for France*, 17 p. http://www.diplomatie.gouv.fr/fr/IMG/pdf/science-diplomacy-for-france-2013_cle83c9d2.pdf

Ministry of Higher Education and Research. 2010. *National Research and Innovation Strategy*, 37 p.

———. 2013. *France Europe 2020—A Strategic Agenda for Research, Technology Transfer and Innovation*, 96 p.

National Research Council. 1999. *The Pervasive Role of Science, Technology, and Health in Foreign Policy: Imperatives for the Department of State*. Washington, DC: The National Academies Press. 124 p. http://www.nap.edu/catalog.php?record_id=9688

National Science Board. 2008. *International Science and Engineering Partnerships: A Priority for US Foreign Policy and our Nation's Innovation Enterprise*, 38 p. http://www.nsf.gov/pubs/2008/nsb084/index.jsp

Roy, S. 2010. *Le positionnement international de la Fraunhofer*. Note of the Service for Science and Technology of the Embassy of France in Germany, March 10.

———. 2012. *Partenariat à l'international des différents organismes de recherche et agences de financement allemands*. Note of the Service for Science and Technology of the Embassy of France in Germany, January 9.

Royal Society and American Association for the Advancement of Science. 2010. *New Frontiers in Science Diplomacy: Navigating the Changing Balance of Power?*, 32 p. http://diplomacy.aaas.org/files/New_Frontiers.pdf

Strasser B.J., and F. Joye. 2005. L'atome, l'espace et les molécules: la coopération scientifique internationale comme nouvel outil de la diplomatie helvétique (1951–1969). In Les nouveaux outils de la diplomatie au XXème siècle, *Relations Internationales* 121(January–March): 59–72.

Sunami, A., T. Hamachi, and S. Kitaba. 2013. The Rise of Science and Technology Diplomacy in Japan. *Science & Diplomacy*, March.

The Cabinet. 2010. *The New Growth Strategy: Blueprint for Revitalizing Japan*. http://www.meti.go.jp/english/policy/economy/growth/report20100618.pdf

Wisard, F. 2010. Le réseau suisse des conseillers scientifiques et technologiques de 1990 à la création de swissnex. *Politorbis* 49: 25–39.

Science in the Vanguard of Diplomacy

5

When political relations between two countries are fraught, or even non-existent, could science be in the vanguard of diplomacy? What this chapter endeavors to show is that particular situations exist in which this question is answered to the affirmative—exchanges between researchers can represent a special relation and, sometimes, even the only form of dialogue between countries sharing awkward relations, or not officially communicating any longer. Scientific relations will therefore be a substitute to usual diplomacy. However, in the very different context of spaces that escape national sovereignties (such as Antarctica) or that are in the process of supranational integration (the European Union), science also provides evidence of its capacity to open up the way to diplomacy. It is to this "science for diplomacy" that this fifth chapter is devoted.

5.1 The Facilitating Role of Science Between Countries Sharing Awkward Political Relations

The capacity that scientific relations have to alleviate international tensions and to facilitate the normalization of inter-state relations has found many examples throughout history, be it during the Cold War or during the ensuing era (de Cerreño and Keynan 1998).[1] The foreign policy of the USA shows several examples where signing scientific agreements was put forward so as to resume bilateral ties which had slackened or even broken.

---

[1] See for instance de Cerreño A. L. C. and A. Keynan (eds.) (1998), *Scientific Cooperation, State Conflict. The Roles of Scientists in Mitigating International Discord*, Annals of the New York Academy of Sciences, 866.

© Springer International Publishing AG 2017
P.-B. Ruffini, *Science and Diplomacy*, Science, Technology and Innovation Studies,
DOI 10.1007/978-3-319-55104-3_5

5.1.1   Of the Usage of Science in US Diplomacy

In 1961, science and technology were mobilized to give a new orientation to the relations between the US and Japan, which had so far been confined to security questions. While receiving the Japanese Prime Minister of the time, President Kennedy launched a US-Japanese committee on scientific cooperation. This innovatory initiative partook of the will to restore a dialogue between the intellectual communities of both countries, which the war had silenced. More than half a century later, programmes for scientific cooperation between both countries are still being implemented by the National Science Foundation. Beyond that example, however, it is during the Cold War, and then at the turn of the century, in the relations with the Muslim world that numerous manifestations of the role played by science in the American foreign policy will be found.

5.1.1.1   The Relations Between the US and the Soviet Union

The Cold War "was a time of highly effective use of science diplomacy to build bridges and connections despite the existence of great political tensions" (Turekian and Neureiter 2012). It is common knowledge that at that time, scientific exchanges between civilian researchers from the USSR and the US were never interrupted (in the same way as trade relations) and were made possible only because the authorities of each country issued the necessary visas. During the 1972 US-Soviet summit, Richard Nixon and Leonid Brejnev signed five major agreements for cooperation in science and technology. Three more agreements were signed the following year. Throughout the whole of the 1970s, scientific cooperation between both countries was flourishing, and hardly altered in 1982 when the US decided to suspend those as a way of protesting against the treatment inflicted onto Jewish scientists as well as onto dissenting physicist Sakharov (Doel and Wang 2002). When it came to the question of nuclear technology, scientific organizations often served as informal media for communicating between both countries. The National Academy of Sciences in the US set up the Committee on International Security and Arms Control (CISAC) in 1980 which from 1981 shared regular relations with the Soviet Academy of Sciences. These annual bilateral meetings—which have gone on to this day with the Russian Academy of Sciences—can be praised for paving the way, in the early 1980s, to the dialogue between Presidents Reagan and Gorbachev on security issues.[2]

---

[2]The role of CISAC extended to China from 1988 and to India from 1999, in order to address technical and potentially sensitive questions such as international security, arms control and disarmament. Its action originates from informal diplomacy (Track II Diplomacy) and is based on the direct interaction between scientists within an international consortium of Academies of Sciences, thus allowing the Committee to have direct ties with heads of states, parliamentarians and military authorities from many countries.

5.1.1.2 The Relations Between the US and China

The first visit of a US president to the People's Republic of China took place in February of 1972. This was a highly significant event diplomatically. It was the first time since the establishment of a communist regime in China that those countries with such antagonistic political orientations started coming closer to each other. During their meeting in the Shanghai communiqué while expressing their will to settle their differences in a peaceful way, Richard Nixon and Mao Zedong came to the agreement to make science an area of cooperation.[3]

The relation between the US and China proves significant here because scientific cooperation came before the process of normalizing diplomatic relations and was at the same time its catalyst (Ratchford 1998). Underpinned by some shared mistrust towards the Soviet Union—although it was for different motives—the relations between the two countries gradually started becoming warmer in 1969. Among the leverage used on the American side to prepare for this turning point was the dispatch to China of renowned American scientists and American Nobel prize winners of Chinese origin C.N. Yang and T.D. Lee. Those even went to meet Mao Zedong and Zhou Enlai, and pleaded with the local political elite to consider the importance of international cooperation for scientific development (Suttmeier 2010a). Between 1972, when the meeting between Nixon and Mao took place, and 1979, when diplomatic relations were set up between the two countries, researchers exchanged with one another on a regular basis. All throughout that period of time, the USA expanded scientific initiatives in order to increase their political and commercial influences, by providing launch services for a telecommunications geostationary satellite, ships for seismic observation, a terrestrial station as part of the Landsat programme for observing the Earth from space, or a synchrotron for particle physics (Whitesides 2010). That period was that of "unofficial exchanges, where, in the absence of formal diplomatic relations, scientists also served as 'diplomats' and shapers of professional elite opinion about China" (Xiaoming 2003). During this period, "scientific, technological, and political factors were thus mixed together" in a way that was particularly representative of scientific diplomacy during the Cold War (Suttmeier 2010a, p. 23).

When diplomatic relations were established in 1979, the two countries signed an agreement for science and technology cooperation that set up a framework within which many more specialized agreements and protocols were concluded in various fields like agriculture, space, energy, Earth science, as well as a significant agreement on students and scholars exchanges (*ibid.*, pp. 27–28). This series of

---

[3]"Nixon and Kissinger wanted to offer something concrete and substantial to the Chinese, going beyond political change that was at the core of their visit. In the remarkable Shanghai communiqué signed at the end of the visit, they included science as a domain of future cooperation between the US and China" (Turekian and Neureiter 2012).

agreements involving American technical agencies stimulated staff exchanges and enabled joint research to be set off. Thanks to it, the US became China's prime partner in the 1980s in terms of formal agreements signed in the fields of science and technology. From the American viewpoint, the whole of the decade which followed the restoring of diplomatic relations was marked by the primate of foreign policy considerations aiming to carry out science and technology exchanges. What motivated China above all was the prospect of getting access to American technology (Xiaoming 2003).

In the following years, and despite the interruption which came after the events at Tiananmen in 1989,[4] this science and technology agreement remained solid anyhow, whereas on other issues, like those of trade relations or questions of ownership, sometimes acute differences between both countries arose. Once these matters were sorted out, the 1990s and the 2000s witnessed Sino-American science and technology relations coming to maturity.

5.1.1.3 The Relations Between the US and the Muslim World

In the 2000s, the relations between the US and some Muslim countries, reputed to be hostile, provide other examples of the facilitating role of science for diplomacy. Remarkable examples can be found of the soft power of science, which, for Americans, according to Norman Neureiter, is defined as "an active way of reaching out to the Muslim world in an area where we know they admire us" (Hsu 2011). This argument has widely been used by the US Department of State: "U.S. S&T capability remains one of the most admired aspects of American society around the world, and this is particularly true in predominantly Muslim countries. Public opinion polling indicates that people view American science and technology more favorably than American products, our education system, or even our freedom and democracy".[5] This viewpoint was embodied in a strategy of recapture (*Muslim S&T outreach strategy*) and culminated in its public expression thanks to President Obama's "Cairo speech" (Box 5.1).

---

[4]Following the events at Tiananmen, the Bush administration suspended top-level political visits. The bilateral science and technology commission did not meet. But the exchanges between researchers continued anyhow—in 1990, all 11 top-level Chinese delegations which went to the US were scientific delegations (Xiaoming 2003).

[5]A statement by J. Miotke, Deputy Assistant Secretary for Science, Space, and Health, Bureau of Oceans, Environment, and Science (U.S. Department of State) at the hearing before the sub-committee for research and scientific education (committee of science and technology), House of Representatives, 110th Congress, second version, 2 April 2008.

Box 5.1 "A New Beginning", speech delivered by United States President Barrack Obama on 4 June 2009 at Cairo University

"... *On science and technology, we will launch a new fund to support technological development in Muslim-majority countries, and to help transfer ideas to the marketplace so they can create jobs. We will open centers of scientific excellence in Africa, the Middle East and Southeast Asia, and appoint new Science Envoys to collaborate on programs that develop new sources of energy, create green jobs, digitize records, clean water, and grow new crops. And today I am announcing a new global effort with the Organization of the Islamic Conference to eradicate polio. And we will also expand partnerships with Muslim communities to promote child and maternal health ...*"

The most original announcement of this speech was the appointment of science envoys to Muslim-majority countries. It was quickly followed up. The first three envoys chosen in 2009 were Bruce Alberts (biophysicist and biochemist, twice President of the National Academy of Sciences), Elias Zerhouni (Algerian-American radiology and biomedical engineering specialist, former director of the National Institutes of Health) and Ahmed Zouheil Zewail (Egyptian-American, Nobel prize laureate in chemistry in 1999). The second crew designated in 2012 consisted of Rita Colwell (infectious disease specialist, former director of the National Science Foundation), Gebisa Ejeta (agronomist, winner of the 2009 World Food Prize) and Alice Gast (chemical engineering specialist, former Vice President for research at the Massachusetts Institute of Technology). Bernard Amadei (civil engineering professor at the University of Colorado, founder of the NGO Engineers Without Borders-USA), Susan Hockfield (neuroscientist, past president of the Massachusetts Institute of Technology) and Barbara Schaal (biologist at Washington University St. Louis, Vice President of the National Academy of Sciences) in 2012 constituted the third group of science envoys. Four others have taken over in January 2015: Peter Hotez (specialist in tropical medicine, Baylor University), Jane Lubchenco (marine biologist, Oregon State University), Arun Majumdar (energy and climate issues specialist, Stanford University), and Geraldine Richmond (chemistry professor at University of Oregon).

Since 2009, these 13 envoys visited more than 20 countries. According to official statements, they travelled as individuals and reported to the White House, the State Department and the American scientific community. Based on personal contacts made by these renowned science envoys, this inexpensive program was generally well received in the countries visited. But it was noted that its effectiveness over the long term depended crucially on the ability to transform these personal contacts in sustainable interactions between research institutions, along with appropriate financial means.

Many agreements were signed in the 2000s, with Algeria, Morocco, Jordan, Kazakhstan and Azerbaijan. Libya and the US signed a bilateral agreement in 2008, which was the first since the official resuming of relations between the two countries in 2004. This agreement was made possible with the return of Tripoli to the international community and with their decision to relinquish their use of weapons of mass destruction, and made provision for cooperation in the fields of public health, water resources and research in the atmosphere.

Relations with Iran are particularly complex. In 1980, the US ceased to have formal relation with Iran. However, despite a major disagreement on the issue of nuclear energy and despite the application of economic sanctions from 2006 to 2015, the scientific communities of both countries have never stopped exchanging. They even have strengthened them—an agreement between the Academies of Science of both countries was concluded in the early 2000s and found expression throughout the decade with about 20 bilateral research seminars (Hsu 2011).

5.1.2   Questions About Science for Diplomacy

All the examples that have been given take the same turn—when political and diplomatic relations between countries are fraught, scientific exchanges and dialogues between researchers permit the maintenance of ties. By not keeping researchers from organizing meetings, or even by facilitating their cooperation, countries engaged in a logic of opposition or of ideological confrontation tacitly reveal their intentions to come out of such situations. And so is science a prelude to the normalization of diplomatic relations, and it is this "softening" role which undoubtedly makes it one of the most remarkable achievements of its soft power.

But these examples bear another common point—they all relate to the US, a country where scientific cooperation has often been put forward in contexts of awkward diplomatic relations. This bias in the analysis cannot be avoided. The existence of conflicts, whether open or latent, is a necessary condition for expressing that particular form of science diplomacy. It is therefore no surprise that the country most exposed to significant international tensions, in times past and even nowadays, should more than others find science diplomacy of great use. The USA is the world's greatest power, and this status, which was acquired after the First World War and which has never been outdone to this day, placed them in the vanguard of the great conflicts of the second half of the twentieth century. After the Second World War, the Cold War represented 40 years of ideological confrontation with the Soviet Union. From the 1950s onwards, this split was coupled with a secondary fight, which opposed the US to communist China. Once the Cold War was over, a new split appeared at the turn of the millennium with the rise of Muslim fundamentalism and of the terrorist organizations asserting their relation to it, here again placing the US in the international forefront.

But there is more to it than being powerful and exposed for a country to be a candidate for the use of science in a strategy of reducing tensions, it also has to be scientifically convincing. The USA does meet this second prerequisite. As the first economic and military power, the US are also first in the field of scientific research.

Though with relative hindsight, their leadership in this field gives them legitimacy to stress science in international relations. This position makes them a much sought-after partner by less developed countries, which benefit from cooperating in the wake of the American research apparatus.

Which lessons can be drawn from these examples for a good understanding of "science for diplomacy"? What should be noted first is that it is politicians who give the starting signal. It is not the scientists who decide to make diplomatic relations warmer. It is at the top-level that it is decided whether the time has come or not. In order to start warming relations, and not to go too fast towards normalization, it may prove an advantage to place science on the front line of ties regained, or to be regained, and all the more so as science—with its values of sharing, neutrality and universality—augurs well for the positive tonality and the good will that wants to be shown in the new course of diplomatic relations. But two questions arise here.

The first one is that of temporality. When scientific cooperation is put forward to open up the way to diplomacy, can this be durable cooperation? This question is to be addressed because the interests of diplomats and those of researchers are not on the same time scale—diplomats are interested in the immediate benefits derived from the signing of a scientific cooperation agreement, which are all the more visible that this agreement comes during a summit of heads of states and within the framework of a process of normalization widely covered by the media. But for researchers, even if an agreement can show the horizons for new co-operations, an agreement signed at the toplevel can only bear fruit in the middle or in the long terms. It is therefore important to check whether diplomatic advantages in the short term find a correspondence with scientific advantages in the longer term, in the form of an increased mobility of researchers, of the training of PhD students, of new scientific results jointly acquired by the two countries. The hindsight we have to grasp the scientific benefits of the opening to China seems sufficient to answer that question to the affirmative.[6] It is not sufficient, however, as we are writing this, to measure if the promises and the expectations phrased in the Cairo speech were met with any effect. The most favorable scenario in order to avoid that the scientific cooperation being put forward by governments should be mere advertising is that in which exchanges and work relations between researchers from the two countries are pre-existent. This scenario is probably the most frequent. American and Chinese researchers did not await Nixon and Mao's green light to start a dialogue, or even to work together. For every field of academic specialty—which are often very specific—scientific excellence is recognized all around the world and is shared among a small group of researchers who all know one another despite their nationalities. Moreover, a political agreement amplifies the movement more than it creates it, by providing a framework that facilitates the expansion of scientific exchanges. Most of all, it sheds light upon what had remained confined into the circle of specialists. For those in power, turning to science is a judicious move, because scientific cooperation is a domain that is, at first sight, little subject to conflicts and that is very likely to lead to interesting results on both sides.

---

[6]For an overview of the scientific cooperation between the US and China since the 1970s, we direct the reader to Xiaoming (2003) and Suttmeier (2010b).

   This leads to a second question. Are researchers exploited, or are they at risk of being so, when they are encouraged to start or to go further into co-operations involving countries in which rulers have decided to strengthen their ties with? Using the phrase "science for diplomacy" almost comes down to announcing, through the choice of words, the subordination of science to the goals of diplomacy and to suggesting that there exists a diplomatic role expected from researchers. To go even further, could a scientist become, as Jean-Jacques Salomon puts it, the "double of a diplomat, of a soldier or of a spy"[7] (Salomon 1989, p. 321)? This question opens another—a larger one—that of the freedom and of the independence of research. The answer depends mostly on the capacity of the public authority to command the activity of researchers. When a country is at war, members of its scientific community are mobilized, either by choice or by duty, into the military effort. During the Cold War, it cannot be excluded that researchers from the Soviet bloc had been enjoined to work with their counterparts from "sister countries" to reinforce the solidarities within the socialist camp—which would nurture an interesting and original chapter on the history of relations between foreign policy and science. However, elements are lacking to uphold this thesis, but it seems very unlikely that such situations could be found in the post-Cold War era. In the contemporary research systems, it is by incentive and not by way of authority that the public powers can orientate the activity of researchers. It is through the allocation of funds that politicians can have leverage on public research. They have to find the right balance between the respect of the freedom of researchers, who want to "devote themselves to science", that is to the fundamental questions imposed by their discipline, and the expectations of society, which are turned to the production of a new knowledge, applicable to the resolution of concrete problems. In public research, researchers are subordinate to nothing other than the means put to their disposal by public budgets.

   The same applies for the research conducted within an international framework. Researchers are not compelled in any way to enter a cooperation programme through which they can see the ulterior motives of a diplomatic agenda. If they do so, it is because they find some benefits as researchers, such that a major actor in scientific cooperation with Iran summarizes the situation in saying that the issue is not whether a question of influence is at stake, but rather whether there is something to learn in the process.[8] And though it does not fall to them as researchers to judge the diplomatic benefits that their country might gain from such a cooperation, they can be happy, as citizens, to see that their work may contribute to bringing peoples closer, and even to reconciliation. It has been noted that there might exist, in the US,

---

[7]Salomon J.-J. (1989), *Science et politique*, Paris: Economica, p. 321.

[8]It is Glenn E. Schweitzer, director of the bureau for Central Europe and Eurasia of National Academies, and who in 2010 was the Science Diplomacy prize of AAAS "for his outstanding record of achievements in demonstrating the powerful role that high calibre science cooperation can have in building international relations". But this American expert points out what is misleading in the phrase "scientific diplomacy", which spontaneously leads, in his view, to the image of the Ministry of Foreign Affairs, and not to that of the Ministry of Science (Badger 2009).

in Britain and in China, "a very strong sentiment on the part of researchers to be the ambassadors of the interests of their countries" (Ministère de l'enseignement supérieur et de la recherche and Ministère des affaires étrangères 2014, p. 85.). This posture seems to be less emphasized by French or German researchers. Eventually, arguments seem to be lacking to nurture the thesis of the exploitation of researchers in the name of science for diplomacy.

Box 5.2 Archaeology in the French science diplomacy

By creating, in 1945, the Advisory Commission for Archaeological Research Abroad—the "Commission des fouilles"—and placing it under the authority of the Ministry of Foreign Affairs, General de Gaulle signaled the strong link between the world of diplomacy and that of archaeology. In the family of scientists, archaeologists are distinctive for their longstanding proximity with diplomats, sharing with them the same interest in the understanding of territories, peoples and their identities.[9]

Archaeology is one of the incarnations of science diplomacy. Just like mathematics, it is one of the jewels of French research and it is generally acknowledged that excellence makes it a powerful vector of prestige and influence. The international reach of French archeology relies on a strong network consisting of 27 research institutes abroad, including more than 10 devoted in large part to archeology, and five *Ecoles Françaises* (located in Athens, Cairo, Hanoi, Madrid and Rome). About 150 archaeological projects are approved annually and certified by the Advisory Commission and performed in over 60 countries.

Archaeological research is rooted in the territories of cities and country-side where excavations take place. This entails the creation of links with local authorities and populations. Archaeological projects promote transfer of know-how and cooperation. Also, far from being limited to the knowledge of the past, the work of archaeologists acts as a support for cultural, educational and linguistic cooperation. Archaeology provides multiple accesses to a foreign country, and that is why diplomats are particularly interested in it. Their interest is even stronger when national experts are requested by a country to learn about its past and regain its heritage and roots: archeology then becomes the instrument of a political dialogue. This discipline has helped to re-establish links with countries emerging from crisis, such as Iraq, where the Roman site of Peramagron was able to be reopened with the help of the CNRS, and Afghanistan, where the reconstruction of the Archaeological Institute was operated with the help of French expertise.

For a country such as France, archaeology is a way of extending its reach and influence. But it is also more than that: it opens doors to diplomacy.

---

[9]French Ministry of Foreign Affairs and International Development (2014), *France and the Promotion of Archaeology Abroad.*

5.2   The Vanguard Role of Science in the Composition of Regional Spaces

Scientific relations may contribute to strengthening ties between communities of researchers from countries belonging to the same region of the world. This stimulating role of science can manifest itself in a regional space without any political existence of its own: large infrastructures for research—such as CERN in Europe or SESAME in the Middle East—contribute to bringing researchers and peoples closer. But the leading role of science can also lie within the framework of a developing political space, such as that of the European Union. Finally, the polar regions are regional spaces with a special status, but from which many lessons can be drawn for this study. This section is devoted to the analysis of these different examples.

5.2.1   The Large Infrastructures for Research in Fundamental Physics and Scientific Diplomacy: CERN and the SESAME Project

The European Organization for Nuclear Research (CERN) was established in 1954.[10] Dedicated to particle physics, it is the largest research center in the world in this specialty (see Box 5.2). Based on an idea attributed to French physicist Louis de Broglie, CERN was established by 12 European countries so as to revive research in physics in Europe right after the Second World War.[11] This institution was founded for two main reasons. The first was to make it an instrument for strengthening the ties between the nations which had harshly fought against one another during the world conflict. The founding of CERN came within the ideological movement for peace and for the construction of Europe, which is characteristic of that period of time.[12] But in uniting their efforts in a peaceful project in this manner, the founding nations pursued a second goal—creating a research capacity in Europe in fundamental physics that would compete with the US, and that would offer the possibility to researchers from the Old Continent to stay in Europe and to work there (Box 5.3).

---

[10]The "Conseil Européen pour la Recherche Nucléaire", a temporary organ established in 1952, preceded the establishment of the organisation. The initial acronym remained the most used to name it, be it in French or in English.

[11]Belgium, Denmark, the Federal Republic of Germany, France, Greece, Italy, the Netherlands, Norway, Sweden, Switzerland, the United Kingdom, and Yugoslavia.

[12]It is during the European Conference for Culture, held in Lausanne in 1949 by Swiss philosopher Denis de Rougemont, that Louis de Broglie presented his ideas. A resolution by UNESCO in 1950 started the process which was to lead to the creation of CERN 4 years later.

> **Box 5.3 The European Organization for Nuclear Research (CERN)**
> CERN is the world's largest particle physics laboratory. Its various facilities and equipment (including particle accelerators such as the *Large Hadron Collider*) are located in Meyrin near Geneva (Switzerland).
>
> CERN is an intergovernmental organization with 22 member states (21 from Europe, plus Israel) and four associate members. Observers include states and organizations currently involved in CERN programs (the European Commission, India, Japan, the Russian Federation, UNESCO and the USA). Nearly 40 non-member states also participate in CERN programs.
>
> CERN is an iconic institution of *Big Science*. It is at the forefront of scientific excellence in its fields of investigation. At CERN in 2012 the ATLAS and CMS experiments provided evidence of the existence of the Higgs boson particle. CERN is also exemplary because its scientific production transcends national affiliations. It employs about 2400 international staff, mainly dedicated to technical and administrative support tasks. It hosts more than 10,000 scientists and engineers annually who conduct experiments for varying durations. In 2013, 113 different nationalities were represented.

The federating power of science to bring nations and peoples closer was a great success. As such, CERN is an emblematic example of "science for diplomacy", not within the scope of bilateral relations as seen in the previous pages with the example of the American foreign policy, but within a multilateral framework. CERN enabled the restoration of bridges between nations separated by historical fractures. It enabled the first contacts after the war between German and Israeli physicists. It enabled scientists from the East and from the West of Europe to work together: during the Cold War, CERN was the first European organization to sign a contract with the USSR. By collaborating with China, it also enabled researchers from Taiwan and from the People's Republic of China to be part of the same team. India became an observer within the organization. CERN signed an agreement with Pakistan in 1994. All the international activity of CERN has thus shown that scientific ties can go beyond any national, racial or ideological antagonisms.[13]

Could the federative experience of CERN be transposed to other regions of the globe? The SESAME project answered that question in the positive. The SESAME synchrotron (Synchrotron Light for Experimental Science and Application in the

---

[13]The political dimension of CERN should be remembered, substantiated by the writings of historians and the testimonies of physicists who took part in this adventure. Was the establishment of CERN first a "foreign policy initiative", closely linked to the objectives of the foreign policy of the US, according to critical observers (Gsponer and Grinevald 2008)? These authors point out that the creation of CERN had far less to see with the Marshall Plan (1948–1952) than with the "Atoms for Peace" initiative by President Eisenhower in 1953. CERN would have then been "trapped in an Atlantist dynamic", which enabled the US, not a member of the international organisation, to gain access to the results of the research conducted in Europe and to indulge in scientific espionage. Cf. also Gsponer A Grinevald (1984) and Krige (2004).

Middle East) is an installation under construction in Jordan, near Amman, which should be put in service by 2017. It will provide researchers from Middle Eastern countries with a source of high quality X-rays. Originally based on the recycling of a particle accelerator built in Berlin by Germany (*Bessy 1*), the SESAME project was born in 1997 at CERN as a result of collaborations with the Middle East. It was officially launched in 2002 under the auspices of UNESCO, as its illustrious predecessor had been half a century earlier. And just like CERN, SESAME aims to develop research in physics for peaceful purposes. The founding states are Bahrain, Cyprus, Egypt, Iran, Israel, Jordan, Pakistan, the Palestinian National Authority and Turkey[14]—this list illustrates the strength of the message of hope and conciliation that science aims to spread in the world's region.

Could other regions of the world strengthen their relations and coherence thanks to scientific cooperation? In the East African Community, an initiative at the presidential level counts on scientific cooperation to stimulate additional regional growth (Turekian and Neureiter 2012). Another example that gets attention is the scientific cooperation in the Euro-Mediterranean space, which the MISTRALS[15] initiative aims to develop. Launched in 2008, MISTRALS is an international meta-programme of interdisciplinary and systematic research and observation for understanding how the environment of the Mediterranean basin functions. This networking of research agencies and institutes from about 30 countries north and south of the Mediterranean for carrying out interdisciplinary programmes gives scientific consistency to the Euro-Mediterranean space, whereas on the political level, partnerships such as "Euromed"—launched in 1995—or the creation of the Union for the Mediterranean in 2008—which aim to go further in that space—hardly manage to take shape.

5.2.2   Science and the Construction of Europe

What is the role of research in the construction of Europe? In order to answer this question, the first step is to understand the meaning of the European project. We favor the thesis according to which a political union was the ultimate objective of the construction of Europe. Deeply rooted in the view of Europe's "founding fathers", this political end was born in the immediate post-war context—the specter of a third world war had to be dismissed for good, and to do so the European countries had to be united together within a common destiny. However, achieving the politics for Europe was not an easy goal to reach. It remains to this day an unspoken objective. There were some transfers of competencies and abandonments of national sovereignty, but political integration is seldom presented as an objective in itself. It is rather seen as the inescapable way out of the process of the construction of Europe or, more positively, as what would crown and perfect the undertaken

---

[14]Just as CERN, SESAME also includes countries which have a status of observer (there were 16 in 2015).

[15]MISTRALS: Mediterranean Integrated STudies at Regional And Local Scales.

integration, once it has reached one by one several main spheres of economic and social life. Ever since the Treaty of Rome was signed in 1957, policies of integration have spread in different ways. First of all, there was commercial integration with the creation of a customs union (the Common Market) from 1958, and then there was a single market for goods and services (the Single European Act) from 1993. After that, there was monetary integration, with countries adopting a single currency in 1999, and the establishment of the European Central Bank: the Eurozone included 19 countries in 2016. Finally, the European debt crisis, which started in 2008, reopened the debate on the coordination of fiscal policies within the currency zone and embodied the prospect of coercion of the European Union on the budgets of member states, which would mean the end of their budgetary and fiscal sovereignty.

The history of the construction of Europe is therefore that of a slower or faster progression on the way to political integration. The gradual Europeanization of the main economic guidelines (trade, money, future of public budgets . . .) goes along with that of other more specific domains, including public research. Research, which is a jurisdiction shared between member states and the Union, is a domain whose goal was set to become a "European area", through reinforcing cross-border co-operations and the increased mobility of both researchers and students. The fact that the Eighth Framework Programme for Research and Development, called "Horizon 2020" (2013–2020), is accompanied with a budget of roughly 70 billion euros—40% more than what had been granted to the Seventh Framework Programme (2006–2012)—shows the importance given to this domain in order to promote the European project. The establishment of agencies, such as the European Space Agency, as well as the construction of large infrastructures, are also indicative of the process of integration through research.

The promotion of a European space for research goes along with the line of reflection developed throughout this chapter. The European countries use research collectively as they use trade, the single currency and all the sectoral policies for integration—that is, as facilitators to build political Europe. This European policy of research projects itself into an original science diplomacy.

Lacking any proper military force, and therefore without any hard power, Europe as such aims to be influential through some soft power, mainly based on its capacity to set norms—references capable of inspiring the organization choices of other countries. An example of the attraction through norms relates to the organization of higher education degrees following a "bachelor-master-doctorate" schedule and of the European Credit Transfer and Accumulation System (ECTS), which favors student mobility, which were adopted by more than 20 countries outside the European Union. The Union also "sells" its cooperative model for research by letting countries that are outside of the Union[16] into its Framework

---

[16]Associated countries and third party countries.

Programmes by allowing eligibility to Marie Curie fellowships[17] or of Erasmus Mundus programs.[18] All these elements show the diplomacy of influence that is fashionable at the moment. It is in this sense that there is a European science diplomacy: in its relations with the rest of the world, Europe should "attract-cooperate-influence", and its research policy is a tool for promoting its model. The goal is to position the Old Continent as a world's reference for the advancement of knowledge, thus sticking to the Lisbon European Council's watchwords uttered in March 2000: to make the European Union "the most competitive and dynamic knowledge-based economy in the world".

But scientific diplomacy in the European Union is also out of the ordinary for having an internal use, in the sense that it addresses the peoples of European countries. Through its initiatives, its programs and its subsidies for research, the Union also "sells" the European project to its citizens—and particularly its researchers, who are the actors and the first to benefit from the construction of the European research area. By favoring the advent of an integrated European community, the research policy operates for the construction of a political Europe.[19]

5.2.3   Science and Diplomacy in the Polar Regions

The polar regions are a world of scientific investigation of utmost interest for scientists all over the world. The study of the world's climate system can be carried out in exceptional conditions: air bubbles trapped in the ice provide very important information on the climate of the past by going hundreds of thousands of years back in time. The polar regions are also irreplaceable places for observing the atmosphere and the stratosphere, particularly seasonal polar ozone holes. These regions are therefore often visited by researchers, from glaciologists to environmental scientists as well as climatologists, astronomers, and physicists studying terrestrial magnetism or the upper layers of the atmosphere. That is what leads about 4000 researchers in the summer and 1000 in the winter to go to Antarctica for stays of various durations.

The polar regions are also unique because of their special place in the analysis of interactions between geopolitics—and therefore diplomacy—and science. Countries covet them as much as they attract the curiosity of researchers. All the conditions are thus met for scientists and diplomats to work together. But the situations in this regard are a lot different in the Arctic and in Antarctica.

---

[17]These fellowships finance the individual mobility of researchers over a period of 2 years. More than 4000 Marie Curie fellowships were granted between 2007 and 2013 to researchers working in 50 countries and representing more than 90 nationalities.

[18]European programmes for the mobility of students and of teachers and researchers.

[19]One should also remember that European science diplomacy has at its disposal a network of five specialised advisers based outside Europe. Many countries also appoint a science counsellor to their Brussels embassies. Many research institutes and universities have permanent representation in the capital of Europe too.

5.2.3.1 In Antarctica

It is science that motivated man to venture onto Antarctica, this unique and uninhabited continent. The will to explore the last frontiers of the planet and scientific curiosity were the reasons that started the great exploratory voyages to the sixth continent (the "Southern Continent") in the nineteenth century. The North Magnetic Pole had been identified in 1831 and it had been deduced that a South Pole existed, which the scientific communities from England, France, Germany and America soon wanted to pinpoint. "What did these men want? They wanted to satisfy an irrepressible desire for knowledge. They had been led there more by the actual strength of their curiosity, their taste for achievement and risk, than really prompted by the strategy of a state, of a mission or of any trading firm" (Victor and Victor 1992, pp. 51–52). Recapitulating the milestones of the discovery of the Antarctic Peninsula, Paul-Emile Victor brings out the idea of a "collective history" (*ibid.*, p. 37): the happy birth of a continent that was to become—and which still is today—the biggest natural laboratory dedicated to nothing else but science.

Research in Antarctica really took off in the middle of the twentieth century. What triggered this rapid development was the preparation for the International Geophysical Year (1957–1958). Adding to its outstanding scientific results, this collective polar venture provides a good example of how science can open up the way to diplomacy. That International Geophysical Year was right in the middle of the Cold War. To the great surprise of the Western countries, the USSR, which, at that time, "did not take part in any Western activity, had never participated in any international cooperation and, at United Nations summits, systematically opposed any suggestion originating from the Western camp" (*ibid.*, p. 103), joined into the preparatory work. The USSR was thus one of the 12 countries to actively participate in the exploratory work in Antarctica. The Soviets built the *Mirny* station, one of the 60 research stations opened for the International Year. Soon after that, they built the *Vostok* station, where both French and Soviet researchers united their efforts in drilling deep into the polar ice.[20]

The International Geophysical Year was also very influential in the genesis of the Antarctic Treaty, signed in 1959 and which became effective in 1961.[21] The diplomats had to find a new and original language so that the Southern Continent could keep its initial status as being entirely for science, by giving a legal status to the scientific spirit which had prevailed from the first expedition. This treaty made Antarctica a territory reserved to science. Several principles were established: non-militarization, and a ban on nuclear test and on radioactive waste repository. The treaty also put an end to the sovereignty rights claimed by ten countries.[22] Antarctica had become a space beyond national jurisdictions.

---

[20] A first-hand testimony on the cooperation between Russian, French and American researchers is offered by Jean-Robert Petit (2008).

[21] This treaty has no limit in its duration and is renewable by tacit consent. It was initially signed by 12 countries. In 2015, there were 53 states party to it.

[22] Namely, the claims of France over the Adélie Land; of the UK, Chile, Argentina over the Antarctic Peninsula; and of Norway, Australia and New-Zealand over other portions of territory in the name of more or less established historical precedents.

Yet, the Antarctic Treaty had left the issue of its governance unanswered. It had been exercised by a scientific authority named SCAR (Scientific Committee on Antarctic Research), which succeeded the special committee established for the International Geophysical Year.[23] Moreover, the treaty had made no provision for the exploitation of mineral resources, environmental protection or the development of tourism—many questions that would soon become important issues.

These issues were addressed during the Madrid Conference held in May 1991. An additional protocol was added to the initial treaty for the protection of the Antarctic environment, which became effective in 1998 under the name "Antarctic-Environmental Protocol". A fifty-year moratorium was introduced, consecrating Antarctica as "a nature reserve, devoted to peace and science" and banning any mineral exploitation.

Antarctica, a "control continent", as Paul-Emile Victor phrased it, therefore provides a unique example of the capacity of scientific cooperation to influence diplomatic decisions. The precedents of the presence of researchers on the sixth continent and their ability to work together, beyond their national belongings, were arguments that proved strong enough to impose onto this territory the respect of a certain geopolitical order, that of peaceful coexistence and of non-militarization. This soft power induced by science had the ability to put global interests above national ones, which can be a lesson in the search for solutions to achieve peaceful governance in other spaces lying outside any country's jurisdiction, such as the high seas, the sea bed, or the cosmos (Berkman et al. 2011).

5.2.3.2 In the Arctic

The situation of the northern polar region is utterly different. It is a space where nations rival with one another in the delimitation and in the appropriation of the Arctic Shelf. However, science, in its relation to diplomacy, is also very active, though sometimes in a somewhat ambivalent way. Beyond the scientific interests it can represent, the Arctic region is coveted today for its economic potential.

Geography helps understand why the Arctic region can be the scene of strong national rivalries. The lands surrounding the frozen ocean that submerges the North Pole belong to five coastal states—Canada, Denmark, Norway, Russia and the USA. Canada first claimed its right of sovereignty in the beginning of the twentieth century and was soon followed by the four other coastal states. Each of them believed that its claim over this region was indisputable. Soon, a system of sectors was adopted to the advantage of countries adjoining the Polar Circle, which can claim a portion of territory up to the Pole. In 1925, Canada became the first state to extend its borders in that direction. The other bordering countries followed suit. The

---

[23]SCAR is an interdisciplinary authority under the aegis of the International Council for Science (ICSU). It oversees the whole of the research conducted in Antarctica and provides the different organs of management of the Antarctic Treaty with its advice.

United Nations Convention on the Law of the Sea (1982) granted to these five states an exclusive economic zone of 200 nautical miles from their coasts. These countries close to the North Pole can therefore enjoy sovereignty over part of the Arctic Ocean, but its main portion represents an international zone.

A major interest in the Arctic region has developed since the effects of climate change became apparent there. The melting of the ice cap extends fishing reserves. It broadens maritime transport routes—the "Northern Route" enables ships to sail from Northern Europe to North-East Asia in about 20 days, against more than 30 via the traditional route going through the Strait of Gibraltar and the Suez Canal. This is made possible 2–4 months a year due to the ocean thawing out. This melting also facilitates access to the enormous resources of fossil fuels lying under the sea bed. According to the United States Geological Survey (USGS), the Arctic region could contain 30% of the gas resources and 70% of the oil present on the Earth.

The international governance of this region is set by the 1982 Convention. It represents the basis of the relations between all five bordering countries, which are the only ones allowed to extend their sovereignties onto the Arctic Shelf. A United Nations Commission on the Limits of the Continental Shelf examines the requests from bordering countries to extend their exclusive economic zone, which have to prove that their continental shelves extend beyond the limit set by the Convention. Such claims have tended to increase in number since the early 2000s. These interventions are contested by other parties, such as the European Union and several environmental NGOs, which demand that this area be placed under multilateral governance, for the sake of the protection of the environment—the Arctic has become a region subject to open tensions.[24]

It is in this context that the posture of science should be considered. Science is mobilized both for the general interest and for the particular interests of the protagonist nations. Concerning the general interest, science is called upon to shed light on essential questions asked by the international community like the predictable reduction of the ice cap, the pace at which the permafrost is melting, or the exact locations of resource deposits. One example was given in a study commissioned by the Geological Survey of Canada from data provided by Canada, Denmark, Finland, Norway, Russia, Sweden and the United States, which in 2009 led to the publication of the first comprehensive geological atlas of the Arctic. Scientific cooperation also provides objective elements that are useful for discussion among states.

But the results of research and the knowledge of experts are also mobilized in national territorial claims. The bordering countries try to bring scientific evidence to back them. What is mainly at stake is to manage to show that the part of the seabed known as the Lomonosov Ridge is the continuation of their respective

---

[24]These tensions are notably expressed within the Arctic Council, an organisation without any binding force and whose role is mainly about making recommendations to the member states. It is composed of the five bordering countries, of three countries with no direct access to the Arctic Ocean (Finland, Iceland and Sweden), of six indigenous representations and of 12 other countries having the status of observer.

continental shelves. The Russians were the first in 2001 to lodge such a request with the UN ad hoc committee, aiming to prove that this ridge runs on from the Eurasian continent. This request was accompanied with a sovereignty recognition claim over an extensive part of the Arctic, including the Pole. The Danes devote themselves to demonstrate that the ridge is a geological continuation of Greenland. They use the conclusions of two expeditions led between 2006 and 2008 to support their claim.[25] Canada pursues a similar objective and tries to show that the Lomonosov Ridge is part of the North American continental shelf. In 2016, no outcome to these requests had been rendered, awaiting complementary investigations.

On the Arctic issue, science on the one hand and politics and diplomacy on the other combine prominently. Science aims to understand the genesis of this part of the Earth's crust and the geological structures of the seabed, as well as establishing its genealogy. Politics and diplomacy want to define the national rights applicable to this region. Science is called in when debates on the new governance of the Arctic are launched. But it will be difficult to apply solutions found for Antarctica—first, because the Arctic is inhabited (by roughly 300,000 Inuit people) and because the bordering states consider that they have some particular responsibilities there. For the observer of the relation between science and diplomacy, the Arctic region is of prime interest, since science has to support diplomacy in contrasted ways. It can be done positively, by bringing an ever more precise knowledge of the polar region to the stakeholders and providing them with a collaborative decisional mode. However, it can also be done in a more ambiguous manner, by serving the national pleas of states competing with one another (Box 5.4).

Box 5.4 Planting the Russian flag on North Pole seabed

Arctic, August 2, 2007. The nuclear-powered atomic icebreaker Rossiya and the polar research vessel Akademik Fyodorov were cruising the Arctic sea with dozens of Russian scientists on board. Over the North Pole, two MIR Deep Submergence Vehicles carried out their descent under the sea cap. After taking various samples of flora, fauna and marine sediments, one of them planted a 1-m titanium flag in the colors of the Federation of Russia on the seabed, 4261 m deep. The success of this first-ever manned descent to the ocean floor at the North Pole commanded admiration. However, it is the geopolitical significance of this gesture that came to the forefront.

This Russian submarine expedition Arktika 2007 reignited what diplomats call an "irritant" in their language. The Russian flag docked at the North Pole indeed expressed more that the legitimate pride of those who pushed the limits of nature with the conquest of a pristine space. Neil Armstrong or the first winners of the Himalayan peaks did not cause any diplomatic incident

---

(continued)

---

[25]The expeditions LORITA-1 between April and May 2006 and LOMROG during the International Polar Year (2007–2008) were formed in order to support the Danish project.

Box 5.4 (continued)

when they waved their national flag on the stage of their exploit. But the Arctic and its outstanding natural resources at hand involve considerable competing interests. Also, the other littoral countries, led by Canada, did not hide their annoyance and express their disapproval. In this gesture they saw a claim of sovereignty by Russia on the North Pole area. This interpretation was fuelled by the leader of the expedition, academician Artur Chilingarov, an internationally renowned scientist and also member of the majority party in the State Duma and special representative of the Russian President for international cooperation in polar regions, who declared after the expedition: "the Arctic has always been Russian and will be Russian!" (reported by news agency Ria Novosti).

In the following days, by way of appeasement, Russia reiterated its commitment to respect the 1982 United Nations Convention on the Law of the Sea. But even if the anchoring of the Russian flag at the North Pole would have resulted from only the initiative of scientists, it has been overtaken by its symbolic significance. Canada and the other countries around the Arctic Ocean could hardly see in it anything other than the political expression of a territorial claim and a use of science for the interests of foreign policy. But does this episode fit in this chapter? In this case, science was not a facilitator of diplomacy, but rather at the source of a diplomatic tension!

With the study of the "science for diplomacy" dimension, this chapter has been devoted to what is perhaps the most complex aspect of science diplomacy. In the context of bilateral relations between states, the role of science as a substitute and vanguard of diplomacy only applies to particular situations, those of strained or even non-existent political relations between countries. The Cold War but also post-Cold War periods provide many examples featuring the foreign policy of a powerful but also very exposed country, the United States. But one must guard against a too US-centric vision. Science for diplomacy is not expressed solely in the context of external relations of the United States. Various other events are found in other parts of the world. We noted the reconciling role played by CERN in Europe after the war and the one that could be expected of SESAME project in the Middle East. We also did a reading of the European Research Area as a facilitator of the political integration of Europe. And, turning to polar research, we demonstrated the ability of scientific cooperation to direct diplomatic choices concerning the governance of Antarctica. We see that "science for diplomacy" goes far beyond the scope of bilateral relations and is present in cases involving multiple countries. The role of science in multilateral diplomacy will be the last step of our study.

References

Badger, E. 2009. *Science Diplomacy: Trading Frock Coats for Lab Coats*. https://psmag.com/
    science-diplomacy-trading-frock-coats-for-lab-coats-6b2cf0434dae
Berkman, P.A., M.A. Lang, D.W.H. Walton, and O.R. Young. 2011. *Science Diplomacy: Science,
    Antarctica, and the Governance of International Spaces*. Washington, DC: Smithsonian
    Institution Scholarly Press. 337 p.
de Cerreño, A.L.C., and A. Keynan, eds. 1998. *Scientific Cooperation, State Conflict. The Roles of
    Scientists in Mitigating International Discord*, Annals of the New York Academy of Sciences,
    866, vii–xiv, 1–280.
Doel, R.E., and Z. Wang. 2002. Science and Technology. In *Encyclopedia of American Foreign
    Policy*, ed. A. De Conde, R. Dean Burns, and F. Logevall, 443–459. New York: Scribner.
Gsponer, A., and J. Grinevald. 1984. *La quadrature du CERN (1984)*. Lausanne: Editions d'En
    Bas. 188 p.
———. 2008. CERN—la physique des particules piégée par l'OTAN. *Horizons et Débats*,
    August.
Hsu, J. 2011. Backdoor Diplomacy: How U.S. Scientists Reach Out to Frenemies.
    innovationnewsdaily, April 8. http://www.innovationnewsdaily.com/196-science-diplomacy-
    soft-power.html
Krige, J. 2004. I.I. Rabi and the Birth of CERN. *Physics Today* 57 (9): 44–48.
Ministère de l'enseignement supérieur et de la recherche et Ministère des affaires étrangères. 2014.
    *La coordination de l'action internationale en matière d'enseignement supérieur et de
    recherche*, 142 p. http://www.ladocumentationfrancaise.fr/rapports-publics/
Petit, J.-R. 2008. *Vostok—Le dernier secret de l'Antarctique*. Paris: Editions Paulsen. 237 p.
Ratchford, T. Jr. 1998. Science, Technology, and U.S. Foreign Relations. *The Bridge* 28(2). http://
    www.nae.edu/Publications/Bridge/EngineeringinChina.aspx
Salomon, J.-J. 1989. *Science et politique*. Paris: Economica. 407 p.
Suttmeier, R.P. 2010a. From Scientific Tourism to Global Partnership (?): Thirty Years of Sino-
    American Relations in Science and Technology. In *Sino-American Relations—Challenges
    Ahead*, ed. Y. Hao, 23–40. Farnham and Burlington: Ashgate Publishing Limited.
———. 2010b. From Cold War Science Diplomacy to Partnering in a Networked World—30
    Years of Sino-US Relations in Science and Technology. *Journal of Science and Technology
    Policy in China* 1 (1): 18–28.
Turekian, V.C., and N.P. Neureiter. 2012. Science and Diplomacy: The Past as Prologue. *Science
    & Diplomacy* 1 (1): 3.
Victor, P.-E., and J.-C. Victor. 1992. *Planète Antarctique—Nouvelle terre des hommes*. Paris:
    Robert Laffont. 273 p.
Whitesides, J.G. 2010. China is Using Scientific Agreements as an Effective Diplomatic Tool—
    Better Diplomacy, Better Science. *China Economic Review*, January 1.
Xiaoming, J. 2003. The China-U.S. Relationship in Science and Technology. Paper presented at
    the conference "China's Emerging Technological Trajectory in the 21st Century", Lally
    School of Management and Technology, Rensselaer Polytechnic Institute, Troy, NY,
    September 4–6, 35 p. http://china-us.uoregon.edu/papers/php

Multilateral Science Diplomacy

6

Multilateral issues did a lot for revealing science diplomacy. The negotiation of international agreements and the often difficult search for consensus between countries on key global issues are among the main tasks of multilateral diplomacy. Scientists take their part in it. They contribute to the definition of the international agenda and they help in finding solutions, thereby supplying the "science in diplomacy" section in the three reference points for interactions among science and diplomacy. This final chapter describes the system of international scientific relations in which global issues arise. It focuses on one of the key issues, that of climate change, in order to examine how science nourishes and supports diplomacy.

6.1 Multilateralization of Institutions, Globalization of Challenges

International scientific relationships take various forms, most of which illustrate the great capacity for self-organization of researchers. Dialogue among peers, circulation of ideas and mobility of people on the global scale take place in professional networks that transcend borders. There are also institutions that allow science to make its voice heard on major social issues. These are private institutions, for example non-governmental organizations such as international academies. These are also intergovernmental organizations with missions which are, in whole or in part, dedicated to the progress of scientific knowledge. The relationship between science and diplomacy in "global public goods" are important within this institutional framework.

6.1.1 Global Challenges, Global Public Goods

The rise of concerns about global issues makes multilateral science diplomacy a timely and consistent approach. Global challenges are those which humanity must

© Springer International Publishing AG 2017
P.-B. Ruffini, *Science and Diplomacy*, Science, Technology and Innovation Studies,
DOI 10.1007/978-3-319-55104-3_6

handle due to risks threatening its survival. The perception of these risks has intensified during the twentieth century. Security as a global issue has been perceived as a risk and faced since the appearance of nuclear weapons. Environment-related risks, which tend to take center stage, entered the collective consciousness in the last third of twentieth century. Threats to human health, which are as old as mankind itself, have been addressed by the international community of states since the creation of the UN system (World Health Organization, Food and Agriculture Organization. . .). Global issues challenge knowledge in different ways. Science is called upon for analyzing causes and effects, thanks to approaches that are often multidisciplinary ones. These great challenges appeal to the community of states to find solutions that are developed and discussed in a multilateral setting.

The concept of "global public good" has emerged in the analysis of global issues. This results from applying the concept of the public good—theorized by experts in public economics in the 1950s. A public good is characterized by the fact that everyone can access it (*non-excludability* property) and can enjoy it without depriving other consumers of that good (*non-rivalry* property). Economics textbooks traditionally cite street lighting or national defense as examples of public goods. The essential question relates to the conditions in which such goods can be produced. Knowing that their consumption cannot be easily individualized and therefore charged to users, public goods are not produced by the private sector but supported by the state (this is an example of "market failure"). By expanding this definition, a global public good is a good "with non-rivalry and non-exclusion characteristics, not only between individuals within a country, but also between people from different countries" (Thoyer 2002). Examples include improving air quality or eradicating a contagious disease: they verify the specific characteristics of public goods and are not related to any particular country, but apply to mankind as a whole.

The global public good concept has flourished in the world of international organizations. The reasons are easy to understand: "To produce and preserve these global public goods, it is necessary that the states in the world cooperate and agree: they must find common solutions to the institutional, economic and political problem posed by the collective identification of global public goods that must be supplied and to provide the means to achieve this" (*ibid.*). Global public goods open the way toward diplomacy: countries need to understand one another and to cooperate in order to face these global challenges in the best way, and negotiations are required to reach agreements or compromises on priority actions and funding. The involvement of NGOs, companies, and representatives of populations and territories is sought in this approach, which illustrates the inclusive new pathways followed by diplomacy at the multilateral level.

Global issues challenge diplomacy, but they ask questions of science first. Many of the challenges that humanity faces are rooted in science and guided by technological choices. A first consequence arises in the sphere of science, the globalization of which is enhanced under the influence of these global issues. But science is also challenged in its relationship with the surrounding world. Policy makers as well as public opinion expect the scientific community to provide in a plain language the state of the art of knowledge on threats to the human species and its environment, without concealing areas of uncertainty or insufficient evidence. As an extension of their role as whistleblowers, scientists are increasingly called upon to put their expertise at the service of public decision-making.

6.1.2 Science Internationals

6.1.2.1 Networks of Academies of Science

Academies bring together scholars, writers and artists acknowledged by their peers. They were historically constituted on a national basis. The first academies of science appeared in the seventeenth century on the Old Continent. The Lincean Academy (*Accademia dei Lincei*) was founded in Italy in 1603. The *Leopoldina*, today *Deutsche Akademie der Wissenschaften Leopoldina*, was founded in 1652 in Berlin under the name of *Academia Naturae Curiosorum*. The *Royal Society of London for the Improvement of Natural Knowledge* was founded in 1660 and the *Académie royale des sciences* in Paris in 1699. The eighteenth century saw the creation in 1724 of the Russian Academy of Sciences. A royal academy of sciences appeared in Sweden in 1739 and another in Denmark in 1742. In 1772 the *Académie impériale et royale des Sciences et Belles-Lettres de Bruxelles* (the *Teresian*) was created. Then, after, the *Academia Real das Ciências* (Lisbon, 1779) and the *Royal Society of Edinburgh* (1783) were founded. Nearly a century later, in 1863, the *National Academy of Sciences* was founded in the United States of America.

Today, a hundred or so countries have an academy of sciences. Almost all these academies include humanities and social sciences within their scope. All are characterized by a strong international component. This is visible first through foreign associate members who are invited to sit in significant though variable proportions, depending on the country. Over one-quarter of the members of the *Leopoldina* are foreigners from 30 different countries. Foreign members represent one-tenth of 1450 members of the *Royal Society*. In France, the *Académie des sciences* has 250 members and 150 foreign associates. The *National Academy of Sciences* has about 2200 members, among which 400 foreign associates. The international dimension is also evident through the bilateral agreements signed by academies. For example, the French *Académie des sciences* concluded 43 agreements with counterpart entities worldwide. But most significant for our purposes are multilateral agreements, which allow national academies to network.

The *Inter Academy Panel* (IAP) is *par excellence* the global network of science academies. With 107 members (104 national academies and 3 networks of academies), it also unites as observers 13 scientific groupings or regional networks of national academies. This network was founded in 1993 and has its headquarters in Trieste (Italy). The *Inter Academy Council* (IAC) is another international network, which was established in 2000 and is based in Amsterdam. It brings together academies of science of 16 countries and works in a complementary way with five international networks of academies.

Adding to these networks with a worldwide coverage, there are groupings of academies formed on a regional basis. Academies of European countries are organized into two complementary networks: *All European Academies* (ALLEA), a federation of 54 academies from 42 countries belonging to the Council of Europe, which was founded in 1994 and is headquartered at the *Leopoldina* in Berlin; and the *European Academies' Science Advisory Council* (EASAC), a grouping of academies of countries of the European Union, which was founded in 2001 and is also based in Berlin. Academies of sciences of the Mediterranean area have teamed up since 2010 in the *Euro-Mediterranean Academic Network* (EMAN), headquartered in Rome at the Lincean Academy. Academies of sciences of Asian countries have gathered into two networks: the *Federation of Asian Scientific Academies and Societies* (FASAS) since 1984, and the *Association of Academies of Sciences in Asia* (AASA) since 2000, the latter involving academies of countries from the Middle to Far East, including Russia. We can add other organizations to this list such as the *Caribbean Scientific Union* (CCC), the *Network of African Science Academies* (NASAC), the *Inter American Network of Academies of Sciences* (IANAS) or the *Network of Academies of Science in the Islamic Countries* (NASIC).

Finally, there are international groupings of academies set up by discipline, such as the *Inter Academy Medical Panel* (IAMP), which since 2000 has brought together over 70 medical academies or medical sections of national academies of sciences.

6.1.2.2 International Academies

International academies are distinguished from international networks of academies in the sense that their members are individuals and not national institutions. Two distinct approaches explain the formation of these academies. One is geographical or geo-economic. The first known international academy is *TWAS—the World Academy of Science for the Advancement of Science in Developing Countries*,[1] which brings together more than 1000 members from 90 different countries: scientists from developing countries—who account for 80% of members—as well as prominent researchers from Northern countries having contributed to the advancement of science in the South. This international academy was founded in 1983 and its primary purpose is to promote science and research in the developing world. The *Latin American Academy of Science* is another example. It was founded in 1983 in memory of Simon Bolivar and is located in Caracas. It brings together over 200 members, almost all Latin Americans. Similar to it in Europe, the *Academia Europaea* was founded in 1988, is based in London and displays over 2000 members representing all scientific disciplines. The *Islamic World Academy of*

---

[1]TWAS—whose name was *Third World Association* until 2004—is headquartered in Trieste and receives a significant portion of its funding from the Italian government. UNESCO is responsible for the management of funds and personnel.

Sciences (IAS), founded in Amman in 1986, gathers for its part a hundred researchers from around thirty countries.

Discipline-based membership is the other approach for creating international academies. The *International Academy of Astronautics* meets this criterion. It was founded in Stockholm in 1960 and has over 1,000 members, who are researchers and engineers from 60 different countries interested in promoting the development of space and aeronautics science. It works closely with the *International Astronautical Federation*, also based in Paris. The *International Academy of Comparative Law*, founded in The Hague in 1924, or the *International Academy of the History of Science*, founded in 1927 and based in Paris, are other examples of international academies built around disciplinary interests.

6.1.2.3 Other Non-Governmental Organizations

We find many other science-based non-governmental organizations in the family of Science Internationals. "International unions", such as the *International Mathematical Union* and the *International Geographical Union* are part of it: there are some 30 of these unions, which contribute significantly to the promotion of science in society by organizing "world years". Another non-governmental organization is the *Pugwash Conferences on Science and World Affairs*, already mentioned in the previous chapter, and whose action in favor of the peaceful use of nuclear energy is a strong landmark in the history of science diplomacy. Finally, one cannot fail to mention here the *International Council for Science* (ICSU[2]). In 2015, this non-governmental organization brought together 122 national members (academies of science, other national scientific institutions) and 31 international scientific unions. Through the interdisciplinary research programs that it promotes on great issues of science in society, ICSU is a symbolic incarnation of the ability of the scientific community to make its voice heard on the world stage.

All these Science Internationals have in common their ability to provide insights on science issues that relate closely to society. They are involved through their expertise in the policy-making process, quite often with respect to issues of multilateral diplomacy. All national academies have the independence and scientific legitimacy required to provide expertise, to report on the state of knowledge, to provide advice and take part in public debates, whether in response to the request of public authorities or on their own initiative.[3] Expertise and consultancy activities are also claimed by international networks of academies and by international

---

[2]The *International Council of Scientific Unions* was created in 1931. The *International Council for Science* has retained this original acronym.

[3]Here is one example: "On the eve of the G8 summit of 2011, academies of the G8 + 5 and the academy of Senegal (invited country) met on 24 and 25 March in Paris for a working session. Following a proposal from the French Academy of sciences, this meeting addressed two major themes: *Water and health* and *Scientific education for global development*. This led to the drafting of a joint statement of the conclusions and recommendations of this group session, which was given to heads of state and government of the G8 countries" (Institut de France-Académie des sciences 2012, p. 27).

academies. Most of these networks and entities were created during the seventies and eighties in the time of growing society's perception of the potential risks of global challenges. For example, bringing a science perspective on global issues to policymakers and to society is part of the stated goals of the Inter Academy Panel. It prepares and publishes statements expressing the consensus of the scientific community on debated issues: such declarations have been adopted about the growth of the world population, urbanization, human cloning, biosecurity or ocean acidification. The Inter Academy Council's core mission is to provide scientific expertise to governments and international organizations on major global issues. The network ALLEA intends to offer its expertise on all matters of science, research and innovation that arise across Europe. Another European network, EASAC, aims to promote joint work of its members to provide guidance to public decision-makers. The Islamic World Academy of Sciences (IAS) plays an advisory role to the Organization of Islamic Cooperation. And, long before the creation of the Intergovernmental Panel on Climate Chang (IPCC), ICSU showed its ability to anticipate future debates by launching together with the World Meteorological Organization an international research program on the atmosphere (Global Atmospheric Research Program) and by welcoming a research group on trends in ozone (Ozone Trend Panel), all initiatives that would facilitate the creation in 1988 of the IPCC.

6.1.3   Intergovernmental Organizations

Intergovernmental science-based organizations are the other major backbone of the framework within which international scientific relations develop. International conventions require special attention because their preparation and implementation define the practical arena of multilateral scientific diplomacy.

6.1.3.1 Intergovernmental Science-Based Organizations

An international organization is a legal entity governed by public law created by sovereign states which team up through an international treaty in order to cooperate in one or more areas, fixed by its articles. A newly created international organization has headquarters, buildings, permanent staff and management and advisory bodies. Such organizations have mushroomed in the twentieth century. There were 37 in 1909 and more than 350 a century later (Zarka 2011, p. 33).[4] For many international organizations, the key area of intervention is directly related to issues of science and, more generally, to the progress of knowledge. They are privileged places for dialogue between scientists and diplomats. Historically, it is the birth of the International Bureau of Weights and Measures in 1875, which provided the first opportunity for such a dialogue: the International Metre Commission was the

---

[4]In a broader sense, the term "international organization" also applies to non-governmental organizations (NGOs), numbering over 4000.

theatre of this dialogue (Salomon 1989).[5] But above all it is the creation of the United Nations after the Second World War and the establishment of the UN system that boosted the surge of intergovernmental science-based organizations: these specialized entities were created by intergovernmental agreements and their connection with the UN resulted from an agreement with it.

The United Nations Educational, Scientific and Cultural Organization (UNESCO), which was founded in 1945, is the UN figurehead with regard to the promotion and mobilization of knowledge in building peaceful relationships between nations. UNESCO participates in this global effort by developing intellectual and moral solidarity between peoples. Its interventions are numerous and multifaceted. They range from supporting major international scientific programs[6] to raising awareness of science issues in public opinion (organization of "International Years" such as the one dedicated to biodiversity in 2010, organization of the International Decade for Action "Water for life" over the period 2005–2015...), to advising states wishing to build capacity in the field of science and technology.

Within the UN system there are partially science-based intergovernmental organizations such as the Food and Agriculture Organization, the International Labor Organization, the World Health Organization, the World Meteorological Organization, the World Intellectual Property Organization or the International Telecommunication Union. The International Atomic Energy Agency is a specialized institution attached to the United Nations through a special agreement of 1957. Originating in the "Atoms for Peace" speech delivered by President Eisenhower in 1953 to the General Assembly of the United Nations, the IAEA is under the aegis of the latter. Its role is to promote the peaceful use of nuclear energy and to reduce its use for military purposes.

The General Assembly of the United Nations has also established subsidiary organs, some of which deal with science and technology related issues: this is the case of the United Nations Institute for Training and Research (UNITAR), the United Nations Environment Program or the World Food Program.

6.1.3.2 International Conventions and Conferences

International conventions provide a legal framework for activities involving several countries. By ratifying them, states commit themselves to respect obligations they codify. There are hundreds of international conventions, many of them affecting areas where progress in scientific knowledge leads to the search for collective rules: the Convention on the Prevention of Marine Pollution by Dumping of Wastes and Other Matter (1972), the Convention on the Prohibition of the Development, Production, Stockpiling and Use of Chemical Weapons and on their Destruction

---

[5]J.-J. Salomon notes that the role of science is purely instrumental: it is "limited to facilitating, at government level, the establishment of institutional links between specialists from different countries" (Salomon 1, in *Science and Politics, op. cit.*, p. 325).

[6]With initiatives such as the World Water Assessment Program, the Man and the Biosphere Program, the International Basic Sciences Program, or the Management of Social Transformations Program.

(1993) or the Convention on Persistent Organic Pollutants (2001) are some examples.

In many ways, international conventions are at the heart of multilateral science diplomacy. First, their proximity to international organizations is strong. An intergovernmental organization is generally entrusted with the monitoring of one or several international conventions: the International Atomic Energy Agency has been responsible for monitoring the correct implementation of the Treaty on the Non-Proliferation of Nuclear Weapons, following its ratification in 1968. An intergovernmental organization can even be created as a result of an agreement. This is the case of the Organization for the Prohibition of Chemical Weapons, which was awarded the Nobel Peace Prize in 2013, and which role is to dismantle the chemical weapons stockpiles of the 189 countries that have committed themselves to. This organization was established in 1997 under Article 8 of the 1993 Chemical Weapons Convention. Secondly, international conventions are a privileged space for dialogue between scientists and diplomats. The terms of these exchanges, which we will detail below on the example of the Climate Convention, are of the same spirit for all conventions, provided that they are nurtured by a science input.

6.1.4 Science-Policy Interfaces

A smooth application of international conventions requires regular assessments and periodic evaluations of the impact of measures implemented. For that purpose convention secretariats commission expert reports, appoint committees whose members are scientists who sit together with representatives of member countries, diplomats and senior officials. These occur upstream of conferences that bring together all stakeholders states, and which are held according to a well-established frequency, usually annually: these are the "Conferences of the Parties" (COP), which are the places par excellence where multilateral negotiations take place.

Although all conventions with a science-related purposes equip themselves with a scientific board and use outside experts and consultants, far fewer can take advantage of the existence of a "science-policy interface", which can be defined as "social processes which encompass relations between scientists and other actors in the policy process, and which allow for exchanges, co-evolution, and joint construction" (Van den Hove 2007, p. 815). These interfaces bring together panels of experts, very often in relation with the operation of an international convention. This practice has expanded with the rise of global challenges, demonstrating the institutionalization of scientific expertise. The IPCC is without a doubt the best known of these interfaces, to which we will devote a lot of attention later in this chapter.

We may also mention the Intergovernmental Science-Policy Platform on Biodiversity and Ecosystem Services, known by its acronym IPBES, which was formally established in 2012 by 94 governments. Worth to be mentioned also is the High Level Panel of Experts on Food Security and Nutrition (HLPE), set up as a science-policy interface of the Committee on World Food Security, which is an intergovernmental body of the United Nations. Other bodies with a more targeted object can be compared with these authentic interfaces, for example the United Nations World Water Assessment Program (WWAP), under the auspices of UNESCO, or the Cooperative Program for Monitoring and Evaluation of the Long-Range Transmission of Air Pollutants in Europe (EMEP), established under the Convention on Long-Range Trans boundary Air Pollution (Treyer et al. 2012).

Science-policy interfaces are first of all characterized by their mission: supplying regularly negotiation and public decision processes with inventories of available knowledge and contributions to the assessment of different policy options. They are also characterized by the methods they apply: their reports and opinions are the result of a collective expertise process and are grounded on scientifically based knowledge. They thus allow for evidence-based decisions, while making sure not to encroach on the responsibility of policy makers and negotiators. Expertise must be, as the expression goes, "policy-relevant" without being "policy-prescriptive". Scientific credibility and independence of experts are major conditions for such practices to be successful: the most effective interfaces (IPCC, IPBES, HLPE) have an independent governance.

Science-policy interfaces emerged recently and have tended to increase in number. The reasons are easy to understand. In the absence of such structures, each of the opposing parties uses its own experts in the negotiations, and any disagreement in the scientific field feeds political and diplomatic disagreements. However, by pointing on consensual scientific views and removing false controversies, interfaces allow for informed decisions from those who have to decide. These structures are not mere passive information providers, and this fully justifies naming them "interfaces". As will be illustrated below, they are places for dialogue between scientists and government representatives. This constitutes their most significant and innovative feature.

Science internationals, science-based intergovernmental organizations and conventions share the same view: global challenges are today at the top of the agenda of multilateral science diplomacy. The climate issue is a good illustration.

6.2 Science in Climate Diplomacy

Climate is a scientific, political and diplomatic issue. Understanding climate phenomena requires the skills of a variety of specialists in natural sciences: climatologists and meteorologists, atmosphere and ocean physicists, glaciologists, chemists. . . . Humanities and social sciences (economy, geography, sociology and law) are also mobilized to analyze the impact of climate change on society and contribute to the design of policies to reduce its harmful effects. Studying climate requires an interdisciplinary approach and raises a genuine scientific debate. The reality of global warming is generally not questioned, but the role played in it by human activity remains under discussion. For a large majority of scientists, it is primarily man who is responsible for warming, because some gas emissions—

notably carbon dioxide—cause greenhouse effect around the planet.[7] This vision is supported by periodic reviews conducted by the IPCC and it leads the scientific community to maintain pressure on policymakers. But although there is a broad consensus, dissenting voices are also heard, for example those of climate skeptics who dispute that warming is largely anthropogenic and to account for it invoke other factors such as long-term natural cycles involving, for instance, solar activity.

Climate change is also a political issue. Fighting against its controllable causes and negative impacts implies finding compromises between groups with conflicting interests: civil society, NGOs, companies. . ... Finally, climate change fits the bill of being a global issue. Greenhouse gases (GHG), which the finger has been pointed at, diffuse globally within a few days and their impact can be held far from where they are emitted. Climate stability is a global public good and defending it requires reaching cooperative solutions at the global scale. Countries must agree: this is the diplomatic dimension. Climate change has been among the priorities of the international political agenda for the last 30 years and tends to combine all development-related discussions.[8] It is at the heart of multilateral diplomacy.

6.2.1   The Institutional Framework

The international climate saga began in response to the alert launched by scientists and settled within the United Nations system.

6.2.1.1 From the First Scientific Warnings to the Creation of the IPCC

Global warming has been reported by scientists in the nineteenth century. The Swede Svante Arrhenius (Nobel laureate in 1903) drew attention to the warming due to the use of coal. He estimated that average temperatures would increase by 5 °C at the end of the twentieth century. But several decades were necessary for climate science to gain ground. The scientific community had reported the anthropogenic origin of climate change since the 1960s. The first warming scenarios were published in the middle of the seventies. The debate on climate change, its causes and consequences, gained currency in the 1980s. International conferences were held[9] and reports were prepared and developed, especially under the aegis of ICSU as already mentioned. The matter of the hole in the ozone layer retained particular

---

[7]A recent compilation of 11,944 abstracts of scientific articles on climate published between 1991 and 2011 by 29,083 authors showed that 97% of those who state a position on global warming believe in its human origin (Cook et al. 2013).

[8]"All other environmental, development or North-South equity issues, which had composed since the 1990s all the topics raised by sustainable development, have tended to be encompassed by the climate regime, subject to the pace of its progress and geopolitical dynamics" (Dahan and Aykut 2012, p. 31).

[9]See for example the report of the World Meteorological Organization (1986).

attention (Benedick 1998).[10] All these steps have prepared the creation of the IPCC.

The IPCC—the Intergovernmental Panel on Climate Change—was created in 1988 by the World Meteorological Organization (WMO) and the United Nations Environment Programme (UNEP). Its secretariat is based in Geneva at the WMO's headquarters. Open to all members of the UN and WMO, the IPCC's mission "is to assess on a comprehensive, objective, open and transparent basis the scientific, technical and socio-economic information relevant to understanding the scientific basis of risk of human-induced climate change, its potential impacts and options for adaptation and mitigation".[11]

The IPCC is the most well-known body of information brought by the scientific community to policymakers. It does not undertake original research nor does it express recommendations. Its large periodical assessment reports (1990, 1995, 2001, 2007, 2014) "should be neutral with respect to policy, although they may need to deal objectively with scientific, technical and socio-economic factors relevant to the application of particular policies".[12] They rely on a state of the art of existing research, on overall observations accumulated by climate scientists and on results derived by scientifically sound methods. This inventory of available knowledge must report on the points of agreement and disagreement in the scientific community. The goal is to "not to seek an average opinion between experts but a consensual writing of dissent that does not eliminate minority positions" (Hourcade 2009, p. 45). The "Summary for Policymakers" is the most publicized production of the IPCC, which is also developing specific reports on a wide variety of more technical aspects in the context of the implementation of the Convention on climate change.

Box 6.1 Milestones in climate diplomacy

1972—Stockholm: United Nations Conference on the Human Environment, generally regarded as the starting point for the interest of the international community for the climate.

1979—Geneva: World Climate Conference—Launch of the World Climate Program. The conference concluded that, due to rising concentrations of greenhouse gases (GHGs), a greater increase in global temperature than has occurred during recorded human history could occur in the first half of the next century.

1985—Villach: World Climate Conference. Growing awareness of the negative impact of the increase in GHG emissions.

(continued)

---

[10]US diplomat Richard Benedick was one of the architects of the Montreal Protocol on protecting the ozone layer.

[11]Article 2 of the "Principles Governing IPCC Work".

[12]*Ibid.*

Box 6.1 (continued)

1987—The Montreal Protocol on Substances that Deplete the Ozone Layer was signed by 24 countries. Objective: 50% reduction by 1999 of emissions of CFCs (Chlorofluorocarbons).

1988—Toronto Conference on "The Changing Atmosphere: Implications for Global Security". Scientists recommended a 20% reduction of GHG emissions by the year 2005 compared to their 1988 level.

1988—Creation of the Intergovernmental Panel on Climate Change (IPCC), in charge of summarizing knowledge on climate change.

1990—Publication of the first IPCC assessment report.

1992—"Earth Summit" in Rio—Adoption of the United Nations Framework Convention on Climate Change (UNFCCC), which planned to stabilize concentrations of GHG in the atmosphere. The Framework Convention entered into force in 1994.

1995—First Conference of the Parties (COP 1) in Berlin.

1997—Adoption of the Kyoto Protocol. Industrialized countries commit to reduce their emissions of the six main GHG by 5% on average between 2008 and 2012 as compared to their 1990 levels, by implementing effective and appropriate measures. Developing countries are exempted from quantified commitments.

2001—US refusal to ratify the Kyoto Protocol.

2005—Entry into force of the Kyoto Protocol. First meeting of the Parties to the Kyoto Protocol (MOP1).

2006—Publication of the Stern Review on the Economics of Climate Change.

2006—China became the world's largest carbon dioxide emitter in the world, ahead of the United States.

2007—Bali Conference. Participating countries failed to reach an agreement on quantified objectives of emission reductions. They only agreed on the idea of a global process, bringing together rich and developing countries.

2007—The Nobel Peace Prize was jointly awarded to the IPCC and to Albert Arnold Gore Jr., "for their efforts to build up and disseminate greater knowledge about man-made climate change, and to lay the foundations for the measures that are needed to counteract such change".

2008—Adoption of the European Union climate and energy package "20-20-20 by 2020": to reduce emissions of GHG by 20% by 2020 taking 1990 emissions as the reference; to increase energy efficiency to save 20% of EU energy consumption by 2020; to reach 20% of renewable energy in the total energy consumption in the EU by 2020.

(continued)

Box 6.1 (continued)
 2009—Failure of the Copenhagen Conference (COP 15): the agreement set the goal of limiting the rise in global temperature to 2 °C but was evasive on how to achieve it.
 2015—Paris Conference (COP 21): unanimous agreement to limit the increase in average temperature to 1.5 °C by 2100 compared to pre-industrial levels; a quinquennial review of national plans for reducing GHGs will take place from 2025; each party shall prepare, provide and maintain its successive and nationally determined contributions.

6.2.1.2 The Setting of the International Climate Regime

The international climate regime is a set of "implicit or explicit principles, norms, rules and decision-making procedures around which actors' expectations converge in a specific field and which can help behaviors to converge" (Maljean-Dubois and Wemaëre 2010, pp. 18–19). The international climate regime brings consistency into various policies at the international level and the different scales of action. It tends to boost the development of national positions with the aim of promoting the formation of an international consensus (*ibid.*).

The international climate regime has evolved in stages. There were promising developments, but also deadlocked periods. Its history is marked by several major dates (Box 6.1). The publication of the first IPCC assessment report in 1990 gave the signal for a general mobilization of diplomacy. This momentum was first concretized with the United Nations Framework Convention on Climate Change at the "Earth Summit" held in Rio in June 1992. In 2016, 197 countries had signed on. Article 2 of this Convention states the ultimate goal: "stabilization of greenhouse gas concentrations of in the atmosphere at the level that would prevent dangerous anthropogenic interference with the climate system". This convention is weakly binding. It sets the legal and institutional framework in which the signatories express their will to cooperate to counteract the negative effects of climate change and negotiate their commitments to achieve it.

The Climate Convention entered in force in 1994. It is the backbone of the UN climate system. Its role has been decisive for structuring the process of negotiations and moving forward the construction of the international climate regime. Since 1995, all signatories have met annually in a Conference of Parties—the "supreme body" of the Convention according to Article 7—to take stock of progress towards objectives and decide on actions to be undertaken.

The IPCC, whose creation was approved by the General Assembly of the United Nations, is fully integrated into the UN system. Its reports written by scientific experts play a major role in the implementation of the Climate Convention.

6.2.2   Hybridization of Science and Diplomacy

How does scientific knowledge get into policy making? When applied to the climate issue, this general question opens on careful consideration of the role of the IPCC, which is "at the heart of the growing and unique hybridization of the dynamics of science itself and of the political dynamics", which characterizes the international climate regime (Dahan Dalmedico 2007, p. 114). Observing how the IPCC works enables us to understand the interweaving of collective expertise in the policy-making process.

6.2.2.1 Working Methods of the IPCC

The preparation of the assessment reports results from a tremendous amount of collective expertise. This work is organized within three groups in charge of drawing up the most accurate state of knowledge possible on scientific, technical and socioeconomic aspects of climate change: Working Group I focuses from a scientific point of view on the climate system and its evolution; Working Group II assesses the impact of climate change on the biosphere and socio-economic systems; Working Group III examines possible solutions to mitigate climate change.

We consider here more specifically the report prepared by the first working group, dedicated to the state of climate science—this first volume being sometimes incorrectly equated with the overall assessment report. Experts who prepare it are "scientists" in the narrow sense of the word. The steering is provided by the IPCC Bureau, with members elected by the General Assembly and mandated by it. The Bureau's primary task is to select authors. For this purpose a call for applications is widely disseminated. Proposals of experts come from national focal points.[13] The Bureau may also encourage applications. In the selection process the preferred criterion is the scientific reputation of candidates, while complying with minority scientific views. But in its choices of authors the Bureau also ensures to balance national origins and genders. To prepare the scientific part of the Fifth assessment report, which was released in September 2013, 259 lead authors from 39 countries were selected from a 1000 applicants. All of them volunteered.

For the preparation of each chapter, two convening lead authors and ten to fifteen leading authors are designated. Their responsibility is to compile and order the results of the scientific literature, which consists of articles published in international journals with referees and of reports and books which scientific value is proven.[14] Then a review process starts, based on a first draft of the report. In a first stage, expert reviewers are invited to correct and supplement the text. This is an open process: the draft report is available for anyone who is interested in bringing complements and criticism. In all cases, comments must be documented by

---

[13] A national focal point is an institution designated by a signatory country of a convention or a member of an intergovernmental panel to be the key contact.

[14] All these procedures are thoroughly described on the official website of the IPCC.

references to publications or by elements contained in the draft report itself. A total of more than 1000 experts have been involved in the scientific part of the Fifth assessment report which was released in 2013.

In a second stage, governments are asked to comment on the collective scientific expertise which has been conducted. The project is channeled to governments' representatives and to relevant ministries through the national focal points. The Summary for Policymakers, whose role is crucial since it is a kind of summary of the synthesis, is also submitted for examination. The text is also forwarded to NGOs. All written comments are incorporated into the report. The first volume of the Fifth Assessment Report has resulted in nearly 55,000 comments. After this long review, the report must be approved in a plenary session by the general meeting of the IPCC, comprising representatives of member countries of the Climate Convention.

The hybridization nexus between science and diplomacy lies in the introduction of a UN-style approach in the work of the IPCC. This work is based on an "organized tension between science and policy" (Hourcade 2009, p. 45). This can be understood from the way IPCC documents are validated. All reports must be endorsed by the relevant working group and by the plenary session.[15] The summary for policymakers is one of the texts to be approved at the end of a line-by-line discussion, which brings together authors of the text and government representatives. The substance of the text cannot be changed, but the way arguments are ordered and highlighted is discussed. In case of disagreement on a formulation, scientists who authored the report have the last word. Thus, the plenary session does not vote on science, but on its formulation.

The IPCC is a unique experience to date of collective scientific expertise at the service of the most prominent of global challenges. The originality of the process should be emphasized. On the one hand, it does not escape the rules and constraints of scientific expertise in general: the discourse of science is never entirely outside the process that generated its expression, because "expertise means nothing by itself, apart from the contexts in which it is built and mobilized" (Dumoulin et al. 2005, p. 10). The general overviews produced by the IPCC are not independent of the questions which were posed. On the second hand, by asking for comments at three successive stages during its preparation, the IPCC shows its uniqueness: these back and forth exchanges question the linear model of expertise in which "the production of scientific statements by experts comes before the political decision" (Maljean-Dubois and Wemaëre 2010, p. 33). In climate diplomacy, there is not a

---

[15]The vocabulary is precise and subtle. Three levels of endorsement are presented on the IPCC website: (1) "Approval" means that the material has been subjected to detailed line by line discussion and agreement. It is the procedure used for the Summary for Policymakers of the Reports, (2) "Adoption" is a process of section-by-section endorsement. It is used for the Synthesis Report and Overview Chapters of Methodology Reports, and (3) "Acceptance" signifies that the material has not been subject to line-by-line nor section-by-section discussion and agreement, but nevertheless presents a comprehensive, objective and balanced view of the subject matter.

scientific debate on one side and political and diplomatic discussions of the other: in global arenas, scientific facts issued by the IPCC become diplomatic facts.[16]

IPCC reports do not make any recommendation nor do they prescribe any measure: that falls under the responsibility of policymakers. It is not that scientists involved in drafting reports lack interest. But it is important for them to be *policy relevant* but not *policy prescriptive*, as the saying goes. For the expert, all the difficulty is to stay in his or her own role, recalling that in general "an expertise does not express science as such, but the conviction of a particular expert" (Roqueplo 1999, p. 48). The IPCC seeks to strike a bold balance between two opposing worlds: knowledge, which must tell the truth of facts; and politics, guided by interests and power relations. In this confrontation, science "cannot and should not have the last word" (Hourcade et al. 2010, p. 33), keeping in mind that "it is not so much a question of telling the truth of science to governments, rather of seeking further progress on a shared conviction about global environmental risks. It is indeed a deeply political process ..." (Dahan and Aykut 2012, p. 24).

Box 6.2 The IPCC assessment reports on climate change: The physical science basis (Working Group I)

1990: First Report

The first IPCC assessment report stated that GHG emissions related to human activity were increasing and likely to intensify global warming. It predicted a rise in average global temperature of 0.3 °C in the twenty-first century.

This first report, yet very measured, sensitized policy makers to the reality of global warming. It marked the start of international negotiations that led to the Climate Convention (UNFCCC) in 1992.

1995: Second Report

The Second Assessment Report announced an average global warming of 1–3.5 °C and an increase in sea level of 15–95 cm during the twenty-first century. This report played a major role in the development of the Kyoto Protocol (1997), which set concrete targets for reducing GHG emissions by industrialized countries.

2001: Third Report

The third assessment report stated without any doubt the causal link between human activity and the increase in GEG emissions. It evaluated to 0.6 °C the rise in average global temperature since 1861, which was 0.15 °C higher than projected in the 1995 report. It claimed that the average global

(continued)

---

[16]Since the Copenhagen conference (2009), limiting global warming to 2 °C between now and 2100 has become a political and diplomatic goal. But although it is a clear and convenient watchword, this warming threshold is based on a questionable simplification of the diagnosis made by scientists (Aykut and Dahan 2011).

Box 6.2 (continued)
temperature could rise by 1.4–5.8 °C and the sea level could rise by 9–88 cm during the twenty-first century. The assessment of impact loomed large in the third report, which prepared the community of states to the need for adaptation.

2007: Fourth Report

The fourth report was more alarming than the previous ones. It confirmed with a high degree of confidence that global warming is due to human activity. According to various scenarios, it predicted an increase in global temperature of 1.1–6.4 °C compared to the 1980–1999 period, with an average value more surely somewhere between +1.8 and +4 °C. It also forecast an average rise of the sea level between 18 cm and 59 cm by the end of the century.

2013–2014: Fifth Report

The fifth report confirmed unequivocally the impact of human activity on climate. IPCC experts estimated to 95% the probability that the accumulation of anthropogenic GEG has been responsible for the elevation of the Earth's temperature since the mid-twentieth century. GHG emissions in the atmosphere could lead to a temperature increase of between 0.3 °C and 4.8 °C for the period 2081–2100 compared to the period 1986–2005. A temperature rise of less than 2 °C was considered weakly probable. The report reassessed the range of the average rise of the sea level by the end of the century to 26–82 cm.

6.2.2.2 Science and International Climate Negotiations

Science feeds international climate negotiations. The IPCC work is their foundation and sets their pace (Box 6.2). This "science in diplomacy" is manifested in the way concepts and diplomatic representations are formed and expressed. The role of scientists is very marked prior to negotiations, but they stand back when negotiations are in progress.

Preparing Negotiations: The Development of "National Positions"

In view of a negotiation, each government develops or confirms its national position which combines two sets of elements: the belief shared by policymakers as to the reality of climate change and its effects on the one hand, and the objectives that the country would like to reach at the end of the negotiation, on the other. In this way the discourse of science is embedded in the interests of national players: the national position is a political construction. The mandate given to negotiators is a diplomatic position.

The method used by the IPCC for the validation of its reports consists in broad discussion and consensus building, as already mentioned. The summary for policymakers, the paramount importance of which should be emphasized again, is developed by circulating back and forth between the bureau of the IPCC and governments, and then adopted after a line-by-line discussion in a plenary meeting where governments are represented. The IPCC is an intergovernmental construction which has the support of all countries that are members. No country officially distances itself from the approach of the expert panel, nor with the assessment reports and their conclusions which have been validated collectively. Human responsibility in global warming and risks implied in the medium and long run, all this is accepted in all capitals.

But an intellectual harmony does not necessarily lead to a political agreement. This surface unanimity of accepting climate change as mainly manmade cannot mask fundamental disagreements between countries when it comes to deciding how to fight against global warming. Results achieved at annual conferences are often weak. The broad consensus among governments on the IPCC messages is consistent with the definition by those same governments of very different political positions, which at times diverge even to the point of being opposite. This seemingly paradoxical situation is a major dimension of the hybridization between science and politics which characterizes the international climate regime.

An interesting study has illuminated the reasons for this paradox by examining how political decisions on climate are built in the United States and in Germany (Grundmann 2007). The hostile position of the US regarding any binding commitment in the fight against global warming is as opposed to the German position. Green movements are very influential in Germany, but not in the US where by contrast lobbies of fossil energy sources are very active. In each of these countries, the press conveys the official position rather accurately, giving ample room for climate skeptics the United States. The two countries converge on another point: the political grip on the process of expertise.[17]

Taking into account the variety of national interests is crucial when moving from the scientific diagnosis to political and diplomatic proposals. But before contrasting national positions and mapping the "geopolitics of carbon" (Maljean-Dubois and Wemaëre 2010), it should be noted that the 1992 Climate Convention does not place all countries on the same footing. Under the principle of equity, it recognizes "common but differentiated responsibilities". In the wake of international law of development, international environmental law establishes a distinction between the obligations of industrialized countries and developing countries. According to the Climate Convention, "the developed country Parties should take the lead in combating climate change and the adverse effects thereof" (Article 3), in the name of their historical responsibility for cumulative emissions since the preindustrial age. Countries with this special responsibility (mainly developed ones) are listed in the Annex to the Convention. There are now 43 in this group. Differentiating between

---

[17]R. Grundmann notes that the United States and Germany, the parliamentary investigation committees have favored hearings of experts supporting the official line: "In the US we saw how Congressional hearings were instrumentalized in order to avoid binding carbon emissions while in Germany the study commission was set up to legitimize exactly such a policy" (p. 428).

countries is particularly important in terms of efforts to accomplish—financial and other—to achieve the objectives of the Convention. The Kyoto Protocol has taken this principle fully and retained differentiated emission reduction targets by country: commitments only apply to industrialized countries.

The variety of national positions on climate, however, reveals no such simplistic opposition of a "North-South" type. Developing countries do not form a homogenous block. Although they are collectively concerned that industrialized countries take high commitments under the Kyoto Protocol, the most dynamic of them claim a right to industrial development, should it even follow a carbon path, and resent that most developed countries having experienced their growth at a time when environmental concerns were absent could dictate their economic development model. These countries have become primarily responsible for increases in annual GHG emissions: according to estimates of the Global Carbon Project, China accounted for 27% of emissions in 2013 (US: 14%, EU: 10%). But it is also among the developing countries that we find the countries which are most vulnerable to climate change, especially small island countries threatened by sea level rises. Their level of development severely limits resources they can allocate to the fight against adverse effects of climate change.

Major carbon energy producers—including Russia—should be set apart. They benefit from oil and gas revenues and do not make the fight against global warming an overriding imperative. In international arenas, at best they adopt a wait-and-see attitude.

Among developed countries, the contrast is sharp between the United States, which above all wants to avoid burdening their economies by imposing emissions restrictions, and the European Union, led by Germany and UK, which emphasizes the precautionary principle and presses toward strong decisions. Being generally hostile to any form of supranational control, refusing interventionism, and suspicious vis-à-vis regulation and binding environmental treaties, the US has been on a hard diplomatic line (Bush junior administration) or a more wait-and-see one (Obama administration) that is well suited to US oil lobby interests. As the largest GHG emitters of the industrialized world, the United States has not ratified the Kyoto Protocol on the grounds that it provided no emissions reduction commitment for the large emerging countries. But this position does not prevent grassroots initiatives to reduce carbon emissions. Climate (and environmental) concern is firmly seated in American diplomacy,[18] reflecting a desire to influence and control the evolution of this major multilateral issue.

The diplomatic position of the European Union on climate is complex in its development, as member countries need to agree among themselves prior to

---

[18]We noted in Chap. 4 the involvement in this field of *Environment, Science, Technology and Health officers* at US embassies.

international conferences.[19] It emphasizes the precautionary principle and the urgent need to act to limit global warming. Under the influence of Germany, then also of the United Kingdom, the Union acquired a leadership role on the climate issue, in international arenas seeking to convince to move towards more binding commitments. Diplomats speaking on behalf of the European Union have done much for the ratification of the Kyoto Protocol. Adding to its prior sensitivity on environmental issues under the influence of Northern European countries, the proactive stance of the EU is also explained by the political and economic benefits of exercising the leadership in the fight against climate change, and of being be somehow the "environmentalist conscience" of the world.

In the end two lessons emerge about the way national diplomatic positions are built. First, the diversity that characterizes them is not determined by science. It is wrong to believe that scientific truths could automatically be embodied in national policy choices. National positions are based on the same foundation of scientific knowledge (IPCC reports), but their diversity reflect that of national interests. In the end, domestic political considerations make the difference: scientific knowledge, as objective as it strives to be, is far from inspiring homogeneous national choices—quite the contrary. This leaves the field open wide for discussions: the annual meetings of the Climate Convention, roughly speaking, see the meeting of negotiators who say they do not know enough to decide for radically binding policies and others for whom it is past time for action, as the evidence of manmade global warming has been widely established.

Second, and to conclude this development on the cornerstones of national positions, it should be noted that the political differences between countries do not cover the differences of opinion that may exist within the scientific community. Of the latter is not possible to make a national reading. On climate, many researchers use scientific evidence to argue that global warming of recent times is largely of human origin. Others, substantially fewer, challenge those conclusions. But this divide is not related to specific national affiliations. The non-ratification of the Kyoto Protocol by the United States has very little to do with the opinion of the American scientific community. From the 1960s and 1970s American scientists have contributed much to the modern climate science, and they played a major role in the creation of the IPCC (with the English and Swedish researchers, to a lesser extent). But they did not influence the official US position, quite the contrary.[20]

---

[19]Building a European position is not easy. In the definition of the "climate package", the Union has been torn between its historical core, calling for ambitious targets for emissions reduction (France, Germany, Italy, Netherlands, Spain, the United Kingdom) and recent member states wishing to prioritize the short-term competitiveness of their economies (Czech Republic, Hungary, Poland, Slovakia). Adding to this, there are other dividing lines on the share of renewable energies in the energy mix.

[20]Suspecting that the scientists' majority conviction was not going in the direction of his vision of American interests, President Bush Junior in May 2001 commissioned a report by the National Academy of Sciences on the state of climate science. In their response 11 leading scientists confirmed the dominant view of global warming having an essentially human origin. The US President did not yield to these argument show ever. The Kyoto Protocol came into force in 2005 and has been ratified to date by 184 countries, but not by the United States.

Negotiations in progress

The Climate Convention was adopted in 1992. Since then, it has given rise each year to a large conference of the parties. It is in this context that negotiations themselves are taking place. We quickly describe here what a conference of the parties is and focus on the role of scientists, which is secondary at this stage, and that of diplomats, which is prominent.

"A "Climate COP" is a kind of showpiece which lasts 2 weeks, with not only actual negotiations where only official delegates express themselves (the *in*) but also a large series of events involving all those who, in one way or another, are concerned with climate issues; this is the *off*, the true lifeblood of the process" (Dahan and Aykut 2012, p. 18).[21] Hundreds of events are held during the same time period, organized by governments, research centers, think tanks, companies, environmental associations and various lobby groups wishing to take advantage of the gathering to promote their ideas, their scenarios or their interests. Several newspapers are also published daily to report about conference news and events which surround. COP 21 (Paris, 2015), one of the most popular to date, attracted nearly 40,000 people, including 150 heads of State or Government.

At a Climate COP, about 1500 people are involved in negotiations in one capacity or another and to various degrees. The first week is devoted to technical work and the second week is more political with the participation of ministers or heads of state. The scheme adopted is pyramidal, which is common in major UN conferences: "The negotiation climbs up the pyramid by progressing in increasingly restricted groups; the agreement is found on its top; then the way down validates, consolidates and completes the central agreement" (Maljean-Dubois and Wemaëre 2010, p. 255).

In the official work of the COP, we can distinguish three groups of stakeholders:

- Technical experts. They are officials from relevant ministries (environment, economy, agriculture...). These experts are negotiators in essentially technical working groups and commissions; without being scientists in the sense adopted in this book, they rely on discussions on science-based documents;
- Diplomats. They are also negotiators. They are involved in policy discussions that take place after the technical aspects have been processed. The delicate task of writing the points of agreement or compromise belongs to them;
- Scientific experts. They are not negotiators but are present in official delegations in order to give advice whenever necessary.

It is ultimately the job of diplomats to act in the political period of the negotiation. At international conferences, experts are not at the heart of the debate and do not have to be: "Scientists, engineers, economists may be influential as experts, they

---

[21]We borrow from these authors the following description elements.

could greatly contribute to framing the regime, but in the dynamics of an ordinary COP like the one of Durban, they do not play a decisive role during *in* sessions" (Dahan and Aykut 2012, p. 18). Their greatest value is upstream of negotiations (assistance in preparing national positions). Besides their participation in parallel events, their presence at COPs is, however, necessary in official delegations both for supporting negotiators and for enhancing the credibility of the whole process.

How can the contribution of science to climate diplomacy be assessed? As we have seen, the IPCC discourse feeds and structures all the climate debate. The IPCC was created in order "to minimize the risk of strategic manipulation of an unstable knowledge" (Hourcade et al. 2010, p. 21). It has managed to secure acceptance of a high level of consensus in an area characterized by scientific uncertainty. The influence of science on climate diplomacy is important. And insofar as the climate challenge tends to permeate all environmental issues, the position occupied by the work and conclusions of the panel indirectly influences all the diplomacy of environment.

The IPCC is the best developed example of science-policy interface. Not surprisingly, it could be held up as a model to follow: the science-policy interface on biodiversity (IPBES) drew inspiration directly from it. The IPCC, however, is not exempt from criticism. Some aspects of its internal operation raised questions, which led in 2010 to the decision to conduct an independent assessment entrusted to the Inter Academy Council. But more fundamentally, although it denies making recommendations, it is sometimes seen as the body that tends to normalize climate policy. This illustrates the fragility of the border between what is *policy relevant* and what is *policy prescriptive* in scientific expertise.

A part of the IPCC asset is to have managed to make Member States of the Climate Convention sharing in a formal and unanimous way its alerts and warnings on global warming. This success helped to make it a Nobel Prize winner in 2007 (jointly with former Vice President of the United States Al Gore). But how far does this "science in diplomacy" make diplomacy move? What we found in this chapter is something of a paradox. As the desire for peace, how sincere it may be, does not prevent those who govern to consider that sometimes war is necessary or inevitable, the desire to see the world reduce its GHG emissions comes up against the diversity of national interests when it comes to translating this into action. In international negotiations the "preference for the sovereignty and the rejection of a top down approach of sharing" (Hourcade et al. 2010, p. 30) are at work, as countries are very unequally prepared to take their part of the burden. In the end, what the climate record teaches us about science in diplomacy is that the road from scientific consensus to political agreement is a long and winding one.

In this last chapter we have detailed the way in which the discourse of science nourishes and supports diplomatic action. We first set the institutional framework, that of "Science Internationals" such as networks of national academies and international academies, which make their voice heard on global issues and thus contribute to the making of policy decisions. Then we turned to science-based intergovernmental organizations, whose role is essential in the implementation of conventions on global issues. Finally, we dealt with science-policy interfaces, which are presently the most advanced forms of the entry of science into the process of political and diplomatic decision. On the example of climate diplomacy, we have detailed the role played by the IPCC, which represents the most accomplished form of science-policy interface.

This last chapter has allowed us to open our thinking to multilateral science diplomacy, as it was until now largely confined to bilateral relations. As compared to the frame of bilateral diplomacy, in which each country seeks to defend and promote its own interests and values, multilateral diplomacy brings a new dimension: governments would not have maximization of national interests as only concern. In the course of multilateral negotiations and "the long process of learning and transaction", a "transformation of national interests and sometimes their slow re-composition around common interests" (Devin 2012, pp. 24–25) could occur. However, this optimistic reading is only partially found in climate diplomacy: national interests are barely able to be fused into genuine common interests, and while the consensus seems largely reached about climate science, it is far from being so when diplomatic decisions are at stake.

References

Aykut, S.C., and A. Dahan. 2011. Le régime climatique avant et après Copenhague: Sciences, politiques et l'objectif des deux degrés. *Natures, Sciences, Sociétés* 19 (2): 144–157.

Benedick, R. 1998. *Ozone Diplomacy—New Directions in Safeguarding the Planet*. Cambridge: Harvard University Press. 480 p.

Cook, J., et al. 2013. Quantifying the Consensus on Anthropogenic Global Warming in the Scientific Literature. *Environmental Research Letters* 8 (2): 1–7.

Dahan Dalmedico, A. 2007. Le régime climatique entre science, expertise et politique. In *Les modèles du futur—Changement climatique et scénarios économiques: enjeux politiques et économiques*, ed. A. Dahan Dalmedico, 113–139. Paris: La Découverte.

Dahan, A., and S. Aykut. 2012. *De Rio 1992 à Rio 2012. Vingt ans de négociations climatiques: quel bilan? Quel rôle pour l'Europe? Quels futurs?* Paris: Rapport pour le Centre d'analyse stratégique. 190 p.

Devin, G. 2012. Influer au sein des organisations internationales: le défi de la coordination. *Mondes—Les Cahiers du Quai d'Orsay* 9: 15–25.

Dumoulin, L., S. Le Branche, C. Robert, and P. Warin, eds. 2005. *Le recours aux experts, raisons et usages politiques*. Grenoble: Presses Universitaires de. 482 p.

Grundmann, R. 2007. Climate Change and Knowledge Politics. *Environmental Politics* 16 (3): 414–432.

Hourcade, J.-C. 2009. Des liens compliqués entre sciences et politique à propos du Giec. *Projet* 6 (313): 42–47.

Hourcade, J.-C., H. Le Treut, and L. Tubiana. 2010. L'affaire climatique au-delà des contes et légendes. *Projet* 316: 19–33.

Institut de France-Académie des sciences. 2012. *Un an avec l'Académie des sciences 2010–2011*. 40 p.

Maljean-Dubois, S., and M. Wemaëre. 2010. *La diplomatie climatique*. Paris: Editions A. Pedone. 378 p.

Roqueplo, P. 1999. *Entre savoir et décision, l'expertise scientifique*. In collection on "*Sciences en questions*". Paris: Institut national de la recherché agronomique. 111 p.

Salomon, J.-J. 1989. *Science et politique*. Paris: Economica. 407 p.

Thoyer, S. 2002. La montée en puissance de la notion de bien public mondial. In *L'Encyclopédie du développement durable*. http://encyclopedie-dd.org/encyclopedie/gouvernance/la-montee-en-puissance-de-la.html

Treyer, S., R. Bille, L. Chabason, and A. Magnan. 2012. Powerful International Science-Policy Interfaces for Sustainable Development. IDDRI Policy Briefs 6. 4 p.

Van den Hove, S. 2007. A Rationale for Science-Policy Interfaces. *Futures* 39: 807–826.

World Meteorological Organization. 1986. Report of the International Conference on the Assessment of the Role of Carbon Dioxide and of Other Greenhouse Gases in Climate Variations and Associated Impacts, Villach, Austria, 9–15 October 1985, WMO no. 661.

Zarka, J.-C. 2011. *Institutions internationales*. 5th ed. Paris: Ellipses. 168 p.

Conclusion

At the intersection of science and diplomacy, science diplomacy is now recognized by many countries as one of the strengths of their international action and their influence. In this book we strived to identify the forms, understand the reasons for and gauge the stakes of this particular dimension of international relations. As a conclusion, we would like to summarize and try to assess the main achievements of these developments.

Not very long ago, a book written on the relationship between science and diplomacy would not have used the vocabulary that prevails today. The expression "science diplomacy" and the first efforts of analysis and conceptualization were born in the twenty-first century. Yet history bears witness to longstanding ties between science and diplomacy. The great voyages of exploration undertaken by European powers in the eighteenth century were the prehistory of science diplomacy. More recently, between the end of World War II and the early 1990s, the Cold War period abounded with examples where science very much colored foreign policy. In retrospect, this was the time where the foundations of science diplomacy were laid, the incubation period of the concept in some way. Thus, states have long practiced science diplomacy, no doubt without ignoring it, but also without displaying it with today's vocabulary. This is indeed the new element that was the starting point of this book: although the twenty-first century was not the birthplace of science diplomacy, it brought it to light as a claimed and assumed approach of a growing number of countries.

In search of a definition of science diplomacy, we referred to national sovereignty as a starting point: the exercise of diplomacy is an attribute of sovereign nations, and there is no science diplomacy without a direct relation to the interests of governments, in one way or another. This is the case when the action of diplomats promotes cooperation among researchers from different countries, or conversely when international scientific relations facilitate diplomatic relations or serve as its vanguard. This is the case finally when scientific expertise helps policy makers and diplomats to prepare and conduct international negotiations. But we also noted that other practices of research stakeholders (universities, institutes...) fall more diffusely in the overall movement of a country's diplomacy. They contribute to feeding and strengthening its influence on the world stage, and ultimately to serve the national interest. We then identified the challenges of

© Springer International Publishing AG 2017
P.-B. Ruffini, *Science and Diplomacy*, Science, Technology and Innovation Studies,
DOI 10.1007/978-3-319-55104-3

science diplomacy, which we summarized in three watchwords: attraction, cooperation, and influence.

Following the path opened by the Royal Society-AAAS report, we then strived to deepen the analysis of each component of the three-pronged approach of this seminal work. We explored the "diplomacy for science" aspect through a comparative study of the diplomatic apparatus of eleven major countries. We took the measure of the differences between these countries both in the way they design and appropriate the process of science diplomacy and in the resources they allocate to it within their network of embassies. We then turned to the "science for diplomacy" component, and gave several examples of how the foreign policy of a country can put forward science and scientific cooperation to achieve its goals. We also deepened the vanguard role of science in the formation of non-national areas, on the examples of the European Union and of the polar regions. Finally, setting aside the field of bilateral diplomacy, we turned to multilateral diplomacy. We analyzed the peculiar hybridization of science and diplomacy which is at work in climate diplomacy, and this constituted a strong illustration of the "science in diplomacy" aspect.

Many other issues have not been addressed in the necessarily limited size of this book. To conclude, we would like to highlight what remains a central question: what is the real scope, what is the effectiveness of science diplomacy? How much does science help diplomacy? What is the value added by the resources of science in the exercise of soft power? How far, in turn, can diplomacy support science? To answer these questions, we should be able to go deeper by scrutinizing the wide beam of interactions between science and diplomacy, which are summarized in the table below. It would be instructive, for example, to measure the effect on the results of negotiations of the scientific contribution at the time of its preparation. Or, in international cooperation, evaluate what science wins from the facilities offered by agreements negotiated between countries, and measure the impact of the support from science and technology networks at embassies. But with regard to issues where quantification is uneasy, assessments are uncertain and judgments too often steeped in subjectivity. The influence of science on the effectiveness of diplomatic action is more diffuse and more difficult to grasp than, for example, their influence on the development of weapons and therefore the nature of the military.

Science diplomacy and its benefits to stakeholders

	Diplomacy for science	Science for diplomacy	Science in diplomacy
Expression	Intergovernmental agreements on scientific cooperation	Parallel diplomacy (*Track 2 diplomacy*)	Scientific expertise
	Action of scientific and technological networks	*Science Envoys*	Science-policy interfaces
Advantages for diplomacy	Promoting cooperation as mode of relations between states	Support to the normalization of diplomatic relations	Better understanding of global issues
	Influencing through science		Assistance with the preparation of multilateral negotiations

	Diplomacy for science	Science for diplomacy	Science in diplomacy
Advantages for science	Support for the creation of large research infrastructures	Expression of a scientific patriotism	Capability of influence on major societal choices
	Support of diplomatic networks to the internationalization of research (mobility, visas...)	Influence on the governance of international territories	Social recognition of science

In the end, we would like to remind and underline that the strength of the mutual interests it involves gives science diplomacy its best chance to be a sustainable practice. From the diplomat's point of view, science is interesting because of its values, and it makes sense to combine them with diplomatic action. First, science is neutral. It does not take part as such in conflicts between states. In the words of the Director-General of UNESCO, "basic research is complex, expensive and always collective: it is an instrument of peace, in addition to being a factor of development" (Bokova 2010). International scientific relations have the advantage of promoting cooperation as a way to harmonious links between states and peoples. On the other hand, communicating via the channel of science can be seen by the diplomat as a way to moderate international tension and normalize interstate relations. More generally, the soft power of science is one of the vectors of the diplomacy of influence. Second, because it transcends borders and ideologies, science is a universal language. Used to thinking without referring to the national context, researchers are useful allies of diplomats when they negotiate agreements or compromises at a global scale. Environment-related diplomacy offers the best examples of this fruitful dialogue between scientists and diplomats. Scientific expertise helps diplomats to better understand global issues and to efficiently prepare for international negotiations.

Science diplomacy is interesting for the diplomat, but also for the researcher. When, after long and complex negotiations diplomats reach successful outcomes on the creation of large international research infrastructures, science is undoubtedly winning. Also, researchers are generally very willing to be supported by diplomatic networks in their international scientific work, or even ask themselves the direct involvement of embassies to successfully complete their projects. Their pioneering role in territories beyond national borders gives them an ability to influence their governance, as we have seen on the example of the Antarctic continent. Finally, thanks to the scientific expertise and interface structures between science and public policy, science strongly affirms its social utility and its ability to influence major societal choices, including environmental ones. Diplomacy benefits from leaning on science, as does science from leaning on diplomacy. But to ensure that symmetrical advantages exist and increase the effectiveness of the approach of science diplomacy as a whole, it would be needed to gather further observation of concrete situations and to get deeper in the analysis.

Reference

Bokova, I. (2010). Le Monde a besoin de la recherche. *La Tribune de Genève*, October 27.

The manufacturer's authorised representative in the EU is Springer
Nature Customer Service Centre GmbH, Europaplatz 3, 69115 Heidelberg,
Germany. If you have any concerns regarding our products, please
contact ProductSafety@springernature.com

Printed and bound by CPI Group (UK) Ltd, Croydon, CR0 4YY
27/04/2026
02097572-0012